Energy Policies in the European Union

Germany's Ecological Tax Reform

Springer
Berlin
Heidelberg
New York
Barcelona
Hong Kong
London
Milan
Paris
Singapore
Tokyo

P. J. J. Welfens · B. Meyer · W. Pfaffenberger
P. Jasinski · A. Jungmittag

Energy Policies in the European Union

Germany's Ecological Tax Reform

With 21 Figures
and 41 Tables

336.278
E56

Prof. Dr. Paul J. J. Welfens
University of Potsdam
European Institute for International
Economic Relations (EIIW)
August-Bebel-Str. 89
14489 Potsdam
Germany
http://www.euroeiiw.de

Piotr Jasinski
Oxecon-Oxford Economic
Consulting Ltd.
1, Tennyson Lodge
Paradise Square
Oxford, OX1 1DU
GB

Prof. Dr. Bernd Meyer
University of Osnabrück
FB Wirtschaftswissenschaften
Rolandstraße 8
49069 Osnabrück
Germany

Dr. Andre Jungmittag
University of Potsdam
European Institute for International
Economic Relations (EIIW)
August-Bebel-Str. 89
14489 Potsdam
Germany

Prof. Dr. Wolfgang Pfaffenberger
Bremer Energie Institut
Fahrenheitsstraße 8
28359 Bremen
Germany

ISBN 3-540-41652-8 Springer-Verlag Berlin Heidelberg New York

Library of Congress Cataloging-in-Publication Data applied for
Die Deutsche Bibliothek – CIP-Einheitsaufnahme
Energy Policies in the European Union: Germany's Ecological Tax Reform; with 41 tables / P. J. J. Welfens ... – Berlin; Heidelberg; New York; Barcelona; Hong Kong; London; Milan; Paris; Singapore; Tokyo: Springer, 2001
ISBN 3-540-41652-8

This work is subject to copyright. All rights are reserved, whether the whole or part of the material is concerned, specifically the rights of translation, reprinting, reuse of illustrations, recitation, broadcasting, reproduction on microfilm or in any other way, and storage in data banks. Duplication of this publication or parts thereof is permitted only under the provisions of the German Copyright Law of September 9, 1965, in its current version, and permission for use must always be obtained from Springer-Verlag. Violations are liable for prosecution under the German Copyright Law.

Springer-Verlag Berlin Heidelberg New York
ein Unternehmen der BertelsmannSpringer Science+Business Media GmbH

© Springer-Verlag Berlin · Heidelberg 2001
Printed in Germany

The use of general descriptive names, registered names, trademarks, etc. in this publication does not imply, even in the absence of a specific statement, that such names are exempt from the relevant protective laws and regulations and therefore free for general use.

Hardcover-Design: Erich Kirchner, Heidelberg

SPIN 10734384 43/2202-5 4 3 2 1 0 – Printed on acid-free paper

Preface

Following the Earth Summit of Rio the Kyoto Protocol was adopted by 160 countries in 1997. Global warming is considered to be a serious threat to the stability of the world climate and to economic prosperity in some regions. Since most emissions stem from the use of energy, the greenhouse topic made energy policy an important policy domain – after the oil price shocks of the 1970s had phased out in the late 1980s, energy policy was considered a core policy field only in some OECD countries. The German government adopted a phased five-stage ecological tax reform in 1990 when fuel prices were raised by six Pfennigs. The government has announced step three of another six Pfennigs in 2001, but could reverse its policy under the impression of massive oil price hikes and public protests in 2000. The government has pointed out that postponing the next steps of the ecological tax reform risks raising labor costs and unemployment, respectively. Ecological tax revenues had been earmarked to finance social security and to maintain contribution rates below 20% of gross wages.

In the following study we analyze the basic challenges of an ecological tax reform in Germany but also the problems of phasing out nuclear energy. Many skeptical voices have been raised arguing that an ecological tax reform will reduce growth and undermine employment growth. The most important innovation which we present here is a bottom-up macroeconomic model which for the first time incorporates an R&D capital stock so that we can focus on the policy option of devoting part of the ecological tax revenue to financing higher R&D. Higher research and development expenditures relative to GDP are necessary for high wage countries in Europe if they are to restore full employment in the post-Cold War environment – with many low wage countries from eastern Europe entering European and other markets. We show that one could improve energy reduction targets while even increasing economic growth and employment. Our analysis allows one to look into real tax reform aspects but also into pricing plus employment effects. At the bottom line our approach emphasizes the crucial role of innovation, structural adjustment and policy innovations for coping with global warming in a way which is compatible with growth and full employment. We maintain that the analysis presented here for Germany is valid for other OECD countries as well.

This book is an extended version of a study which we prepared in 1999 for the European Parliament, DG IV. By presenting this enlarged version we hope to stimulate the discussion of ecological tax reform both within the Economics profession and among policymakers. We gratefully acknowledge editorial assistance by Ralf Wiegert, Potsdam and Tim Yarling, Tokyo.

Potsdam and Washington, DC, August 2000
Prof. Dr. Paul J.J. Welfens, Jean Monnet Chair in European Economic Integration

University Libraries
Carnegie Mellon University
Pittsburgh PA 15213-3890

Contents

Executive Summary

Energy policy is an element of infrastructure policy and thus is important for competitiveness and growth, at the same time it is a crucial element of environmental policy since the generation and use of fossil and nuclear fuels goes along with negative national and international external effects. Following the EU electricity liberalization initiative Germany has not chosen to implement the minimum gradual liberalization required by the EU, rather it has fully liberalized the electricity market in April 1998 which will lead to falling electricity prices and industry restructuring in a more competitive European market. Since natural gas is an important input – with a competitive edge vis-à-vis alternative inputs - in electricity generation the liberalization of the gas market in the EU initiated by the European Commission will reinforce the liberalization of the energy market. Energy generation and use are in turn key elements for several emissions, most notably CO_2 and SO_2. These gaseous emissions naturally create transboundary pollution problems, other international aspects of ecological tax reforms concern the competitiveness of the tradable goods industry and trade in energy resources and electricity. Moreover there will be effects on international capital markets to the extent that there will be relocation of energy intensive industries or intensified merger and acquisition activities in the energy sector or in energy-intensive industries facing sharper price and cost competition.

The development of nuclear energy use in Germany depends strongly on political decisions to come. Existing nuclear power plants at present are competitive, but building new ones is rather unlikely under conditions of competition. Only by government intervention could the nuclear option be maintained. So if the phase-out of nuclear energy is to come, the question will be how to replace the present electricity generation of nuclear power plants.

As electricity and gas markets are being opened and liberalised throughout Europe, part of the substitute could consist of electricity imports. Also the number of gas fired power plants will increase. Nevertheless, a fast phase-out of nuclear energy means destroying a considerable amount of capital. The assessment of this capital depends on the assessment of the replacement of nuclear power plants. That depends e.g. on future energy prices and especially on the pace of phase-out. A scenario of replacing nuclear power to a great extent by gas power plants offers favourable chances at today's prices and costs. At the same time such a strategy requires natural gas to be available in sharply rising amounts without changes of price. The last condition seems not very probable in view of the natural gas market structure.

A mixed strategy for the replacement by a mixture of natural gas and coal is more acceptable considering price and cost aspects, but raises larger problems with climate policy, because much higher emissions of greenhouse gases are related to such a strategy.

A phase-out neutral to climate issues, on the other hand, raises the costs of phase-out. Because either other economic sectors have to contribute more to greenhouse gas reduction, or the electricity sector itself has to increase the use of natural gas to be able to reduce greenhouse gases. This means refraining from creating a great part of domestic value added in the area of coal production.

A higher contribution of renewable energy cannot solve the problem: The use of RES (renewable energy sources), like the use of nuclear power, is not directly related to the emission of greenhouse gases. If nuclear energy could be replaced by renewable energy sources, the climate balance would remain neutral in the long-term. But this would not be a contribution to the reduction of greenhouse gases.

There are large potentials for combined heat and power generation in Germany, which are not yet exploited. Decentralised plants that provide cheap heat could make an important contribution to power and heat generation in the future. Under today's economic conditions there are still big obstacles for the introduction of such plants, because the market does not remunerate the environmental advantage of co-generation.

When phasing out nuclear energy, energy policy faces the problem of combining environmental soundness and economic soundness in the framework of a global economy as well as the stability of political conditions. Considerable tasks of structural change are related to this. A social pact would be useful which envisages a prolonged operation period for certain nuclear power plants in connection with increased support for RES and CHP (combined heat and power); stimulating energy saving also is important.

For nuclear waste, the main options are reprocessing or ultimate waste disposal. If today ultimate waste disposal is cheaper, it would be economically logical to choose ultimate waste disposal, which for the time being will be long-term intermediate storage. Later, technically it would still be possible to opt for reprocessing. The German Waste disposal concept at present aims at three sites of final storage, for heat developing waste it is the salt dome in Gorleben, which is still being examined. Waste disposal being a considerable cost factor, it is estimated that the dismantling costs of a reactor amount to about 15-20% of the construction costs. The disposal of nuclear waste accounts for approx. 60% compared to 40% for the operational stage.

Taking up some doubts in the literature about the double dividend hypothesis we propose within a modified approach an innovation augmented ecological tax reform: From the theoretical perspective of an optimal allocation approach it makes sense to internalize negative external effects from emissions and positive effects from R&D while using ecological tax revenues to finance R&D promotion and to cut high labour taxes which cause considerable distortions – lower taxes also will promote higher employment. Higher R&D expenditures allow a faster accumulation of the R&D capital stock, which plays a positive role in macroeconomic and sectoral development. The increased R&D capital stock contributes to higher growth and thereby eliminates the negative output effect – possibly associated with intensified struggle for income - observed in most simulations on the effects of the standard ecological tax reform. In a macro model based on input-output analysis we find that about 10% of tax revenue from the ecological tax reform should be spent on higher R&D expenditures if a negative output effect is to be avoided in Germany. With respect both to Germany and the EU we argue that an innovation-augmented ecological tax reform is an ideal strategy to combine environmental improvement with higher employment – roughly plus 1 million.

Our study leaves open several questions including those of time lags between R&D accumulation and employment (and output) growth. Moreover, there are some important issues related to the international effects of higher R&D-GDP ratios. A rise in the R&D capital stock might attract higher foreign direct investment inflows, which in turn could go along with a higher investment-GDP ratio and higher factor productivity at large. It is obvious that most EU countries are likely to benefit from adopting an innovation-oriented ecological tax reform, except for those countries whose industries are weakly positioned in technology intensive products and technologies. An open question is whether part of the extra R&D funds should be devoted to specific research projects with a focus on improving energy efficiency.

These and other issues define an agenda for future research. Given high unemployment rates in most EU countries modified ecological tax reform cannot substitute for adequate labour market reforms. Ecological taxes should provide the impulse to reduce the overall tax burden, minimize the dead-weight welfare losses and optimize the allocation of resources. While Schumpetarian ecological tax reform would be useful for Germany and the whole EU such a reform is no substitute for structural reforms in the social security system of Germany and other countries with a traditional pay-as-you-go system. Policymakers in Germany should modify the existing ecological tax reform in accordance with the proposed Schumpeterian elements. In principle Germany's ecological tax reform could – with adequate modifications – also be applied to the EU. Tax harmonization in the field of ecological taxes should be phased in the future.

Perspectives

A Schumpeterian ecological tax reform could not only bring major benefits for Germany but for the whole EU (it also would be applicable in the US, Canada and

Japan). Since in 1997 per capita CO_2 emissions of Belgium (12 tons) and the Netherlands (11.8 t) were somewhat higher than in Germany (10.8 t) the benefits of a Schumpeterian ecological tax reform in these two countries could be even higher than in Germany. With British, Italian and French per capita emissions being slightly lower – that 9.4, 7.4 and 6.2 t, respectively – a Schumpeterian ecological tax reform would also generate considerable benefits in the UK, Italy and France; the latter, of course, faces special problems from its large nuclear power industry. It would be useful to adopt an EU harmonization approach in the field of (Schumpeterian) ecological tax reform, the two main reasons being that some minimum harmonization would avoid distortionary effects on intra-Community trade and foreign direct investment on the one hand, on the other hand taxation of primary energy inputs according to the respective CO_2 intensity hardly is possible without an EU-wide approach. Germany's ecological tax reform should be modified adequately which means both adopting an explicit focus on CO_2 emissions of primary energy and splitting the ecological tax revenue in a way which would not only reduce labour costs but also raise R&D promotion and R&D expenditures, respectively.

The EU should adopt a Schumpeterian ecological tax reform project, which would generate new jobs, strengthen EU competitiveness and stimulate economic growth. Facing the New World of economic globalization an adequate EU action program could considerably contribute to modernizing the EU and to successfully cope with global warming.

Fig. 1: Innovation-oriented Ecological Tax Reform

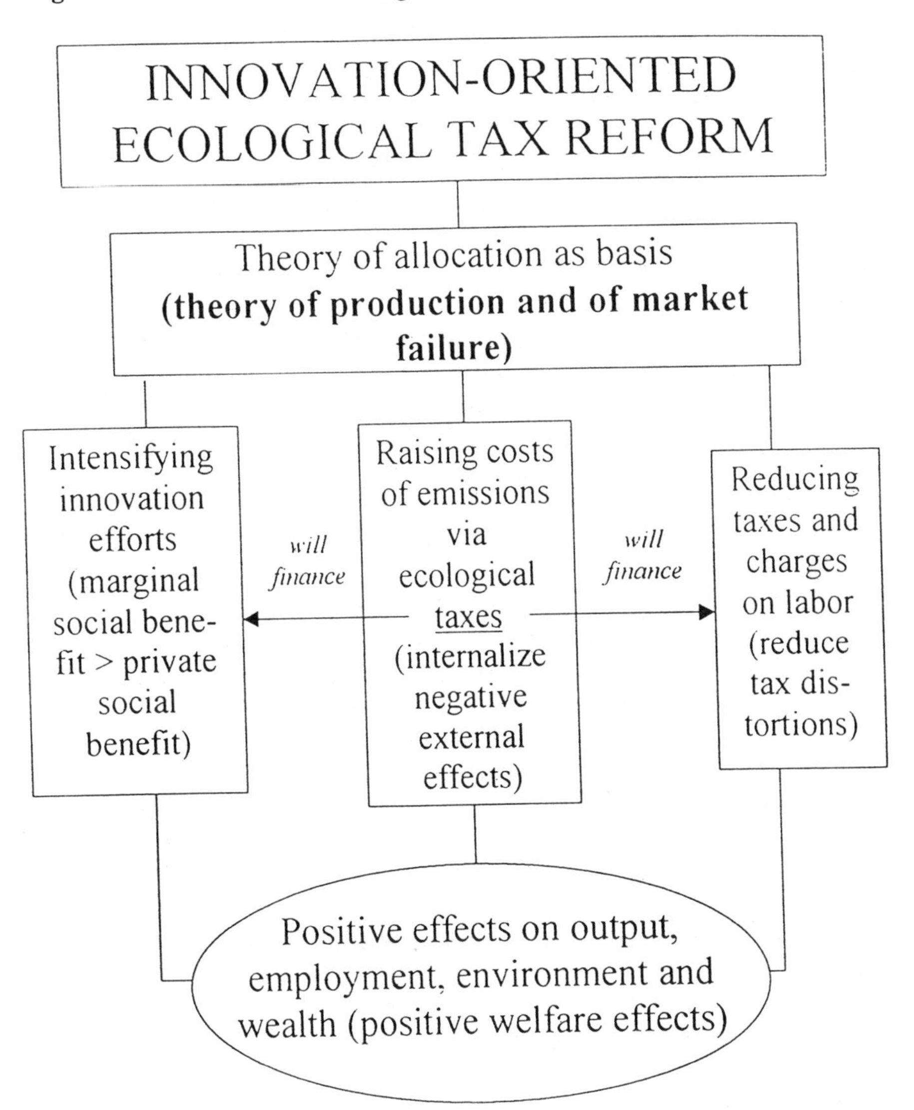

The Method of Simulation

The econometric model gives us an empirically based picture of economic development and its relations to energy consumption and greenhouse gas emissions. The empirical basis for modelling behavioural equations is 1970-96. In a first step of our analysis we make a forecast from now to the year 2010, where various exogenous variables (e.g. world import growth) are fixed at plausible levels. Thereby we get a reference scenario, the so-called business as usual (bau) case. This is dubbed bau case because the forecast and the assumptions lying behind this scenario do not involve policy activities in the field of our interest – here ecological tax reform.

In a second step we produce a further forecast, but now we assume the introduction of a carbon tax and its "compensation" in the sense of revenue-neutrality: In the analysis conducted here is at first a simulation of a simple ecological tax reform with revenue neutrality, i.e. all ecological tax revenue is used for lowering social security contributions. All other assumptions concerning especially the values of the exogenous variables are the same as in the bau scenario. Then further forecasts are made by assuming that a certain share of the ecological tax revenue will be devoted to raising research and development expenditures. This allows us to model a Schumpeterian ecological tax reform, which has an explicit focus on growth-enhancing stimulation of innovation expenditures.

In our discussions of the forecast results we always refer to deviations from the bau forecast. These figures give us all direct and indirect effects, which the carbon tax and the compensation method chosen (e.g. full reduction of social security contributions, or, split of ecological tax revenue for lowering social security contributions and for raising research and development expenditures) will have. E.g. if the forecast result for a Schumpeterian ecological tax reform (SETAR) - with x% of ecological tax revenue devoted to higher R&D expenditures – says in the column for GDP + 1.5% this means that comparing gross domestic product at the end of the forecast period 2010 for SETAR with the business as usual case real output exceeds the figure of the bau case by 1.5%. If a hypothetical SETAR forecast table for nominal wages W says +2% and for price level +0.8%, the result is that at the end of the simulation period the nominal wage is 2% higher than in the bau scenario, while the real wage will be roughly 1.2% above the level in the bau scenario.

Simulation thus allows policymakers to understand which impact the realization of alternative policy options will have. Simulation results from a broad model – including input-output analysis – are econometrically sometimes not as robust as estimation of reduced forms of equation systems, but the broad model has the important advantage of covering complex dynamics of reality in a bottom-up approach and of allowing a closer look at the structural/sectoral changes that are often so crucial for adjustment in the economy, but for policymakers decisions as well.

The model simulation run for one country cannot easily be extended to a group of countries unless national models are linked adequately. Covering several EU countries (or the whole EU-15 or EU plus selected accession countries) is possible in principle but requires international cooperation in a rather complex simulation model. As regards policymaking such simulation analysis can help to avoid serious policy pitfalls and may help to identify the most menu of instruments for a given set of policy objectives.

1. Energy Policy as a Strategic Element of Economic Policy in Dynamic Open Economies

Energy policy is an element of infrastructure policy and thus is important for competitiveness and growth, at the same time it is a crucial element of environmental policy since the generation and use of fossil and nuclear fuels goes along with negative national and international external effects. Following the EU electricity liberalization initiative Germany has not chosen to implement the minimum gradual liberalization required by the EU, rather it fully liberalized the electricity market in April 1998 which will lead to falling electricity prices and industry restructuring in a more competitive European market. Since natural gas is an important input – with a competitive edge vis-à-vis alternative inputs - in electricity generation the liberalization of the gas market in the EU initiated by the European Commission will reinforce the liberalization of the energy market. Energy generation and use are in turnkey elements for several emissions, most notably CO_2 and SO_2. These gaseous emissions naturally create transboundary pollution problems, other international aspects of ecological tax reforms concern competitiveness of the tradable goods industry and trade in energy resources and electricity. Moreover there will be effects on international capital markets to the extent that there will be relocation of energy intensive industries or intensified merger and acquisition activities in the energy sector or in energy-intensive industries facing sharper price and cost competition.

In 1994 the German federal government confirmed its objective of cutting CO_2 emissions by 25-30% between 1987 and 2005. More importantly, the government announced at the COP1 conference in Berlin (1995) that it would reduce CO_2 emissions by 25% over the period 1990-2005. As regards the EU Germany made an important commitment in the European Council of Environment Ministers of March 1997 when the German government agreed to cut its greenhouse gas emissions – that is the three gases CO_2, CH4 and N2O – by ¼ between 1990 and 2010; this commitment was part of the proposal of the European Union for the Kyoto conference in December 1997. The broader emission reduction target concerns the year 2010 which for macroeconomic adjustment – and computer simulations – therefore is the relevant time range. There have been several reports on the reduction of the CO_2 emission target. In our analysis we will focus both on CO_2 emissions and other ecological and economic aspects of energy policy in Germany – including the issue of phasing out nuclear energy. Selected developments in other EU countries also will be considered (including the UK with its liberalized electricity markets whose dynamics offer basic lessons for the EU with its phased-in liberalization).

The Kyoto protocol has called for considerable reduction of CO_2 emissions by all major OECD countries. Raising energy prices in some form thus becomes a natural strategy for many market economies, and a so-called ecological tax reform – including energy/ CO_2 taxes - is one possible option for policymakers. Energy pricing and ecological tax reforms are core elements of sustainable energy policies in industrialized countries. While nuclear power generation has no CO_2 emissions there are serious problems with potential contamination and with nuclear waste management. Clearly, energy policy has not only ecological aspects at the national level, it also concerns cost competitiveness of firms and prices for electricity, gas and heating in private households. Energy policy therefore is a politically and economically sensitive issue.

Economic globalization (WELFENS, 1999a), i.e. the increasing competitive pressure in goods markets has stimulated companies′ activities in cost cutting, including in the field of energy costs. With respect to electricity Germany′s prices were among the highest in the EU in 1997/98 – with east-Germany′s prices being even somewhat higher than those in West Germany (mainly due to high capital allowances in the context of new power stations constructed within the framework of regional monopolies). Scandinavian electricity prices are relatively low in the EU.

In accordance with the Federal Constitutional Court, which had declared the so-called Kohlepfennig – a duty imposed on electricity to subsidize the German coal industry – illegal, the Kohlepfennig, a special charge on households' electricity bills, was eliminated. Instead, subsidies for the hard coal industry were directly disbursed from the federal tax revenues. Government discussed replacing the Kohlepfennig with revenues of DM 4 bill. with a general energy tax, but it turned out that the conservative-liberal coalition government could not get broad support in the Parliament for such fiscal reform. The new government of Germany which took office in autumn 1998 – a coalition government of social democratic and green parties – decided that it would pursue a so-called ecological tax reform which would consist of two elements, namely imposing new energy taxes, while reducing social security contributions; this measure should reduce wage costs and thereby help to raise the demand for labour and to cut the unemployment rate, respectively.

The German government under Chancellor Schröder decided to introduce a phased ecological tax reform, which would raise energy prices. While the experience of the OPEC price shocks of the 1970s gives reasons to expect OECD market economies to react flexibly on the demand and supply side to a new vector of energy prices the plans for an ecological tax reform will bring some additional elements for policymakers and the business community, respectively. An ecological tax reform would bring an additional revenue source; more precisely the revenue ecological taxes would rise. It amounted to roughly 2.5% of GDP in Germany and OECD countries, respectively, in the 1990s. One has to consider to what extent isolated national energy policies and ecological tax reforms are adequate in the framework of the single market and increasing global competition. For the business community a national ecological tax reform implies differential adjustment pressure depending on the energy intensity of technologies and products. Moreover, to the extent that an ecological tax reform allows the reduction of wage costs – here employers´ social security contribution – one will have a temporary cost advantage, which could improve international competitiveness while stimulating employment and thus aggregate domestic demand at the same time. Given high unemployment rates in most continental EU countries in the 1990s an ecological tax reform can only be one element of a much broader strategy to restore full employment. Deregulation and more wage differentiation as well as more mobility are needed in any case (ADDISON/ WELFENS, 1998). For Germany and some other EU countries the second big challenge of modern energy policy concerns options to shift from nuclear energy towards a bigger role for renewable energy sources as well as more conventional alternatives.

An ecological tax reform basically will raise energy prices and thus give incentives – at given technologies - for firms and consumers to save energy by adjusting the input structure on the one hand, and the output mix on the other. At the same time there will be incentives for industry to come up with energy-saving innovations and new technologies. Given the diverging energy intensities of industries, raising energy prices is likely to have different impacts across sectors both in terms of output and employment – in the case of tradable products also in the field of exports. Moreover, for an open economy the changing incentives to import energy and to export energy-intensive products will have effects on the trade balance. The sectoral output effects and the parallel impact on income and normal tax revenues will bring about crucial employment and government revenue effects. It is clear that the revenue effect from imposing higher energy taxes normally is positive for the budget. But how will the economy react if output is falling as a consequence of higher energy prices – an effect typically found in most model simulations on ecological tax reform? Will there be accentuated conflicts over income, increasing long-term budget deficits or major exchange rate effects? The very purpose of the ecological tax reform, namely to improve the environment and to raise employment could be seriously undermined if there were strong adverse secondary effects of an ecological tax reform. An ecological tax reform which would mainly rely on an input substitution effect of the tax reform – namely replacing capital by the now cheaper labour – but would go along with negative long term output effects is unlikely to be sustainable. Hence, to the extent that there are potentially considerable negative output effects of an ecological tax reform, one has to discuss opportunities to mitigate the negative output impact. Here lies a major policy problem for which we will suggest empirically founded solutions for the case of Germany, but our innovative policy proposals should also work in other OECD countries.

2. Phasing out Nuclear Energy and Core Elements of Sustainable Energy Strategy

How to replace the present electricity generation using nuclear power plants in the future? Technically it is no problem to introduce modern huge power stations operating on the basis of natural gas or coal as a substitute. On the other hand, there are certain problems of location related to particular plants, because replacement plants, needing a different infrastructure, sometimes cannot be built at sites where today's nuclear plants are located. In general there are solutions to these problems of location, but they might lead to shifts in the market shares of enterprises or call new suppliers into action. This is a strain for single companies, but that is economically acceptable.

In the context of a European Market national borders do not play the same role as before in electricity supply, because open trade in electricity between the countries of the European Union is intended by the special EU Internal Market Directive. There still remain some transitional problems and unequal structures in different countries, in particular European prices for using the transmission grids are not yet harmonised. But it is only a question of time, just as in Germany transmission prices will evolve that are appropriate for an open market, this will also be the case on a broader European scope.

Bottlenecks in grid capacity can restrict trade for a certain time. At the moment it is not imaginable that the whole production potential of German nuclear power plants could be supplied externally. Nevertheless, when abandoning nuclear energy, the question is: How large will be the share of inland generation in the future? It has to be taken into account that other countries, outside the EU (Switzerland, Eastern Europe, Russia etc.), could be interested in supplying electricity. Thus, part of the substitute could consist of electricity imports.

In the context of the European Union the reciprocity clause applies in electricity trade. It says that trade can be denied to countries that do not allow trade themselves. It is not very likely that this clause could prevent electricity from being imported as a substitute for nuclear power. It is true that the progress that has been made in market opening varies between different countries of the EU, but it has to be considered that the Directive came into force only recently and that it allows a transitional phase.

Towards non-EU-Countries, and Russia in particular, electricity trade might become more difficult. On the other hand Germany has a considerable interest in steady gas supplies. For natural gas, which has a special meaning as a potential substitute for nuclear energy, a trade war, as a means of impeding electricity imports, is not very likely. However, in the long-term there is no reason to believe in a very high growth of electricity imports from EU-countries: electricity generation costs are mainly determined by the costs of power plant erection and the energy costs. These costs are about the same range throughout most European countries, the market for power plants and energy being global. Thus abandoning nuclear energy does not necessarily lead to a migration of great parts of electricity supplies, as might be suspected.

Replacing nuclear power plants by modern fossil plants economically raises fewer problems, if the remaining time of use can be extended in a way that the erection of substitutes can be handled without problems. On the other hand environmental pollution with emissions of greenhouse gases would increase. That is why other options are being discussed intensely. They especially consist of Combined Heat and Power Generation (co-generation, CHP) and the use of Renewable Energy Sources (RES). Finally, a more efficient use of electricity for many applications can reduce the specific input of electricity so that – disregarding economic growth – a decrease in electricity consumption can be possible. The three options and their potential contributions are being addressed in the following pages.

2.1. The Role of Renewable Energies and Innovative Promotion Regimes

Role of Combined Heat and Power Generation

For improving energy efficiency and the environmental balance of energy supply, the need to reach a higher share of electricity production by CHP-plants has long existed. With CHP, two useful products are created, i.e. heat and power. When the heat created in a power generation process can be used economically, the value added of the process increases. The potential of CHP can be classified by the demand for heat from:

- Industrial enterprises: The demand usually strongly depends on location and is highly concentrated, so that there is a high potential when the generated electricity can also be used. On the other hand this potential depends on a specific type of production and its specific market risks.
- Heating and hot water demand in residential buildings and commercial buildings: This demand typically varies very much, because high demand only occurs in a few winter months and the heat density may be low. Therefore, the structure of demand is different to industrial applications.

The potential can be exploited by district heating (town or district grids). Taking CHP as a substitute for nuclear power plants, for a long term analysis it has to be assumed that CHP has to be as economic as new power plants or heat generation plants.

The technical potential of CHP is large and makes up for about one third of electricity generation (including industrial CHP). Only a small part of this potential has been realised up to now. There are two obstacles to the introduction of CHP:

- A large quantity of capacity on the electricity market; and
- The specific obstacles that emerge from an economic use of heat in competition to e.g. natural gas or individual heating.

In the following we concentrate on the first obstacle, because the conditions on the heat market specifically depend on local and regional factors. Practically the whole stock of fossil power plants will have to be replaced within the next 30 years. After the year 2008, the demand for substitutes increases heavily until the year of 2015, and reaches an annual average rate of 3 GW. In cumulated terms it can be stated that approximately until 2015 30 GW of fossil capacity will have to be replaced, if the same electricity generation has to be provided. This proves a high potential for CHP.

If and to what extent CHP will be a substitute is mainly a question of institutional conditions of the electricity market. These are changing dramatically at the moment due to market opening. At present, the possibilities of an extension of CHP are assessed rather sceptically on account of heavily dropping electricity prices. Nevertheless, the new policy framework also provides many advantages, because electricity prices for buying electricity (up to now high) and for feeding electricity generated from CHP into the grid (up to now more or less low) tend to adjust.

It is possible that technologies based on decentralised heat and power generation will spread more widely in the future and then also small heat potentials could be served by local CHP-plants. It is expected that power can be generated from natural gas in fuel cells, whose size could rather flexibly be adjusted to the particular demand. Also the district demand could be matched that way, so the additional costs for the distribution of heat could remain very low, which could improve the economic balance of the systems. Also, it has to be pointed out that taxing energy sources for heat generation (taxes on natural gas, heating oil) has a negative effect on individual heating systems, because they make less use of the input energy than CHP-systems.

To a certain degree it depends on the political will to implement CHP as a more efficient form of energy use in the future. In that case political measures could lead to a broader dissemination of CHP, as was already the case in some European neighbouring countries (the Netherlands, Denmark), since there still is a gap between the macro-economic advantage of the better energy efficiency of CHP-utilisation on one side and the economic risks of an extension of heat transmission grids with their high investments and the long term financial commitments on the other.

Less energy demand through the combined generation of heat and power also leads to less environmental pollution. However, this emissions advantage depends on the energy sources deployed in power generation. The conversion of all existing CHP-potentials could lower CO_2 emissions of the power sector by up to 20%, if natural gas is used.

Whether this potential can be realised also depends on which other measures are taken for reducing energy consumption in other areas. It is known that improved heat insulation of buildings can reduce the energy consumption for room heating to quite an extent. It can also be assumed that legal measures, already existing for new buildings will also be implemented to a greater extent for old buildings in the future. The reduced demand for heat resulting from this measure also lower the potential of CHP.

Taking all power plants to be shut down within the next 30 years in Germany, it becomes clear that almost the whole present stock will have to be replaced. Abandoning nuclear energy is not a necessary pre-condition to get CHP going in Germany. For the further development of CHP, performing, reasonable, small plants that adjust well to local demand are important. These have to be competitive in the long run compared to large power generation in cheap huge plants.

Role of Renewable Energy Sources

In the future, the use of renewable energy sources could offer the opportunity to replace the energy sources used so far. For power generation, hydropower, wind power, direct use of solar radiation by photovoltaic or other solar power plants as well as biomass can be used in particular. Tab. 1 shows the contribution of renewable energy to the total energy supply and to power generation in Germany. About 2% of total primary energy supply comes from renewable sources, for power generation the share amounts to about 5%, which is mainly hydro and wind power.

The potential of hydropower is more or less being exhausted and a further increase of hydropower possibly leading to conflicts with nature conservation, an increase of the contribution of RES has to come from other sources. Today, the use of renewable energy in power generation is supported by the Electricity Supply Law. Generators of renewable power get a legally fixed minimum price, which is related to the average power price level. The renewable power is fed into the general power system and when paying their bills, consumers automatically pay for the renewable energy, which would otherwise not be economic compared to market prices.

These incentive measures today have to be revised due to the liberalisation of electricity markets. One possibility is a further development of the Electricity Supply Law towards more competition, the implementation of a quota system is another. For the latter, a certain share of power that has to be generated from renewable sources is fixed by the government. Producers, distributors or consumers can be obliged to meet this share. The producers of renewable power receive certificates for the renewable energy generated, which can be sold e.g. on a stock market and are used as a proof for fulfilling the obligation. This model could also be interesting with respect to an international electricity market, since some European countries (The Netherlands, Denmark) have already begun to implement such a system, and cross border trade of green energy as well as certificates could be possible. Also in Germany the discussion about the introduction of a quantity based incentive for RES and also for CHP has started.

Independent of the outcome of this discussion of new stimulation incentives for renewable energy in the electricity market, it can be stated that in the future additional financing for renewable energy will definitely be necessary. For wind the demand for financing depends on the quality of the location. The more wind can be "harvested" at one location, the better the economic balance of generation is at this particular location. The "harvest" depending on the third power of the velocity of the wind, even small differences between locations have a great importance (the velocity of wind being 10% higher, the yield increases by about one third!).

Tab. 1: Contribution Renewable Energy, D 1998

Renewable Energy 1998	PetaJoule
Hydro power (estimated)	59
Wood (1994)	(47)
Sewage sludge, Waste (1994)	(92)
Sewage gas(1994)	(13)
Wind (estimated)	18
Photovoltaic	0.04
Total	284
Total primary energy consumption Total contribution RES approx.	14320 2%
Electricity generation RES 1998	**TWh**
Hydro power	19
Wind approx.	4.6
Waste etc. approx.	2
Biomass approx.	1
Photovoltaic	0.03
Total	26.5
Corresponding approx.	5%

Source: BMWi, Energiedaten '99; Energiebilanz; own calculations; For Power: IWR Münster

The better locations not being available anymore, a further dissemination is only possible by:

- developing locations that show a less favourable average wind velocity, which would lead to higher costs
- installing new, better performing and bigger plants that could increase the yield at existing locations later in time
- exploiting new potential for offshore plants.

The special problem with wind energy is that due to erratic supply the contribution of wind power cannot be planned reliably, so in addition to every wind power station, further power plant capacity has to be available for replacing the capacity of the wind power station, if wind is not available.

The use of biomass for combustion provides interesting possibilities, because here the supply can be planned. Because of the low energy density of usable biomass, the transport effort is relatively high, that is why decentralised plants are more appropriate for this kind of fuels.

Today, the generation of growing proportions of power from renewable energy sources is demanded. This has been taken into account in all our scenarios for alternative time-schedules for closing nuclear power plants. Higher capacities for power plants on the basis of renewable energy are envisaged. One could say that this is only necessary in the case of a phase-out of nuclear energy, but this is not quite true. Renewable energy has such great importance for future energy supply that it has to be developed anyway, no matter which power plants will be used apart from that. Thus they have been included in the scenarios as part of modernising the power plant stock in Germany.

Improvement of Energy Efficiency

Looking at the total energy consumption in Germany from extraction of primary energy sources up to the energy services (collectible energy) finally desired, it becomes obvious that only one third of the total primary energy can be used as collectible energy. Two thirds are lost either in the transformation stages e.g. at electricity production or in the transformation devices of the user itself, i.e. it can not be used for the desired purpose.

Therefore, in the future, an improvement of the exploitation of energy is an important source to achieve the same benefit with less energy input. Finally, it can be checked, how much energy is needed for which areas of use to achieve a certain benefit. Summarised, there are the following possibilities:

- Improvement of energy use in the transformation stage (see also the part about CHP)
- Improvement of energy efficiency at the level of energy users, and finally

- Optimization of energy use ("energy saving").

The potentials for efficiency improvements and energy saving are high in all economic sectors. But in the discussion, often-general energy saving potentials are mixed up with that specific for the electricity sector.

As far as electricity is concerned the following potential for improvement exist:

- Substitution of electricity by non-electricity, e.g. for heat generation
- Improvement of efficiency with electrical appliances and using systems
- Reduction of demand of collectible energy for certain applications.

The efficiency of electricity use is partly related directly to the technology of appliances and partly depends on behaviour. Stand-by-losses, accounting to 20 TWh belong to that. Improvement of technological energy efficiency takes as long as the complete replacement of the stock of appliances will take. On the other hand there are many political possibilities to influence the efficiency standards for technologies of new appliances and to reduce the consumption level in the long-term. Comparing the best available technology with the average installed technology, there are huge energy saving potentials concerning the specific consumption of appliances. Assuming that the population as well as the use of electrical equipment remains constant, a decreasing demand is very likely. However, the unknown in such considerations is the development of the population and of the national product in the future which directly or indirectly determine the future number of electrical appliances and their use. It also depends on whether the consumption in absolute terms will decrease or increase.

At the moment a decline of population is forecast for the time after 2005 due to the age distribution in Germany. However, an important influence on the real development of population is immigration. Given the forthcoming extension of European Union towards Eastern Europe and the high attractiveness of Germany as an immigration country, the question of the immigration to come is of great importance for many economic orders and the resulting patterns of energy consumption. Here, politics concerning the admitted volume of immigration indirectly is also responsible for the consequences for energy and electricity consumption. (OTTE, 1999)

Tab. 2: End Energy and Collectible Energy

	End energy	Losses	Collectible energy	Degree of efficiency
Industry	Mt SKE			%
Process heat	57	23.4	33.6	58.9%
Space heating	9.5	2.8	6.7	70.5%
Mech. Energy	15.7	5.7	10	63.7%
Lighting	1.2	1.1	0.1	8.3%
Communications	1	0.1	0.9	90.0%
Total	84.4	33.1	51.3	60.8%
Transport				
Process heat	0	0	0	0.0%
Space heating	0.3	0.1	0.2	66.7%
Mech. Energy	88.3	72.4	15.9	18.0%
Lighting	0.3	0.3	0	0.0%
Communications	0.3	0	0.3	100.0%
Total	89.2	72.8	16.4	18.4%
Households				
Process heat	13.6	7.3	6.3	46.3%
Space heating	69.4	18.7	50.7	73.1%
Mech. Energy	4.5	2.7	1.8	40.0%
Lighting	1.4	1.3	0.1	7.1%
Communications	1.7	0.5	1.2	70.6%
Total	90.6	30.5	60.1	66.3%
Trade/Services				
Process heat	11.8	6.7	5.1	43.2%
Space heating	27.9	8.1	19.8	71.0%
Mech. Energy	10.2	4.3	5.9	57.8%
Lighting	2.8	2.6	0.2	7.1%
Communications	1.2	0.2	1	83.3%
Total	53.9	21.9	32	59.4%
All Sectors				
Process heat	82.4	37.4	45	54.6%
Space heating	107.1	29.7	77.4	72.3%
Mech. Energy	118.7	85.1	33.6	28.3%
Lighting	5.7	5.3	0.4	7.0%
Communications	4.2	0.8	3.4	81.0%
Total	318.1	158.3	159.8	50.2%

Source: RWE Energiebilanz 1995

2.2. Alternative Time-Schedules for Closing Nuclear Power Plants

The effects different development paths of the power plant stock in Germany might have are here examined. A simulation model is used that enables us to compare different futures with each other. We concentrate on the question, what effect does the availability or non-availability of nuclear power plants has? Different assumptions about the future are examined with additional sensitivity calculations regarding the development of electricity demand as well as prices and costs.

Assuming a lifetime of 40 years, the electricity generation from nuclear energy up to 2030 has to be replaced by newer plants. The decision cannot be made by only considering these plants. The demand for electricity probably will change in the future. We here take the future demand for electricity as given and constant. Within limits, the existing plants can be used differently. On the other hand they also have an age distribution and within the period up to 2030 many of these old plants will have to be replaced by new ones. It is logical that considerations about the use of such plants have to be related to the future fate of nuclear power plants. From that point of view, we consider the total stock of power plants existing today and how it will have to be developed in the future, to provide the same production as today.

When abandoning nuclear energy, the existing plants will have to be used in a different way than today and new plants will have to be built to replace the generation lost. To be able to analyse all these problems in an understandable way and thus simulate different future scenarios, we built up the total power plant stock of Germany in detail as the stock of a "German joint stock company" that makes decisions about the distribution and development of this power plant stock for the future, according to homogenous criteria. From a macro-economic point of view this is appropriate. Beyond that, by focusing on part of the total power system (i.e. today's level of demand) this allows an optimal adjustment of the future power plant stock to this demand. Thus we can leave out many questions, such as the question of the future role of power imports or the future role of improvement of energy efficiency, because all these issues have an effect on the growth of demand, on which we will not focus.

Fig. 2 shows the general perspective for power generation in three different scenarios. In total the biggest part of today's generation until 2030 will have to come from new plants replacing existing fossil power plants and nuclear power plants. The contribution of renewable energy sources has been pre-fixed by us.

Fig. 2: Electricity Generation (Scenarios)

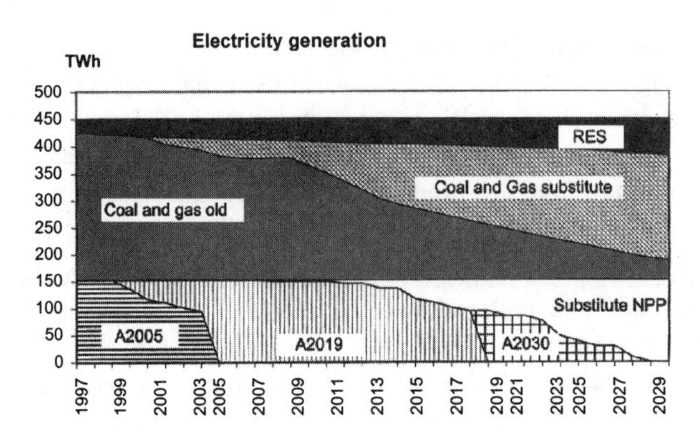

We calculated three different scenarios of nuclear energy phase-out and assessed the macro-economic effects of nuclear energy on the basis of the difference between the three cases. The first case describes the phase-out of nuclear power plants according to their technical end for an assumed lifetime of 40 years. The second case describes the phase-out of nuclear energy until 2019 and the third case a fast phase-out until the year 2005. Due to the age distribution of "other" power plants there are a lot of considerable changes in this area during that period of time. Thus, the scenarios also differ in the design of other power plants for a longer transitional period. The end of the scenarios (2030) has been chosen, because until then all existing nuclear power plants will have to be replaced anyway, because of reaching the end of their technical lifetime. Thus, here the scenarios meet again. But the mix of plants in this particular year is not necessarily the same, because different decisions about the replacement of power plants can be made according to different conditions in the course of time.

Definition of Scenarios

In public discussion different variations for the settlement of nuclear energy are being discussed. Our variation phase-out in 2005 is based on considerations by Bündnis 90/Die Grünen. Here the proposal was to put a time limit of 25 years to the operating permits in retrospect, and additionally to set a deadline year for abandoning nuclear energy. In our scenario this has been taken into account by gradually taking all nuclear power plants off-grid that reach an age of 25 years and by closing down all other nuclear power plants by 2004.

In the medium-term scenario until 2019 it is being assumed that all nuclear power plants can reach a maximum lifetime of 40 years, if this is possible up to 2019. All nuclear power plants have to be closed down by that year at the latest, anyway.

Phase-out in 2005: with a transitional phase of one year after coming into force of the legal amendment the operating permits of each nuclear power plant are limited to 25 years and additionally at the end of 2004 all remaining nuclear power plants have to close down.

Phase-out in 2019: The operating permits remain basically unlimited. However, all nuclear power plants run 40 years maximum and will be closed down in the year 2019 at the latest.

Phase-out in 2029 (reference scenario): The operating permits remain unlimited. For every nuclear power plant a maximum lifetime of 40 years is calculated. A gradual decline in the use of existing nuclear power plants until the end of 2028 will be the result.

A Simple Estimation

A simple assessment of the value of nuclear energy can be achieved by calculating the extra kilowatt-hours produced in the reference scenario with the average extra price, which would apply in the phase-out case. The fuel and operating costs of electricity generation of depreciated nuclear power plants are about 0.04 DM/kWh. The extra costs for power from other (new) power plants lie between 0.025 and 0.05 DM/kWh, depending on the input energy source and type of power plant.

For a phase-out in 2005 there are total extra costs of 69 billion DM, assuming specific extra costs of 0.025 DM/kWh, and for the year 2019 of 12 billion DM. As will be shown later, this is (on the basis of today's fuel costs) an over-estimation of the real extra costs. There are two main reasons for that:

1. For this calculation we pretend that all kWh still to be generated by nuclear power plants, have to be replaced by other new power plants. This must not necessarily be the case, depending on the use of capacity of existing plants.
2. In the case of phase-out, the substitution plants will also be fully depreciated after a certain period and then they only will have to recover the fuel and operating costs. Since the period under consideration is longer than 30 years, while power plants normally are depreciated within 20 years, this is a relevant factor for our analysis.

A methodologically better result can be achieved by looking at the development of the total power plant stock and also by considering different assumptions about the development of fuel prices. Here, a complex dynamic calculation is necessary, in which the contribution of every single power plant can be adjusted to each price-cost-situation and also the substitutional power plants can be planned on the grounds of price-cost-considerations. Our analysis is based on such a model.

Further a simple calculation as this one does not take into account the effects on the economy and the environment. Macro-economic effects, as well as environmental effects can only be analysed with the help of modelling a more complex scheme.

Methodology

In connection with the capacity of power plants for public supply, all scenarios are based on the consideration that the existing free capacity will not remain at the same level in the future. In total the capacity will decrease so that in the medium-term the available net capacity of 90 GW will remain. This reduction of capacity is not going to be homogenous for all types of power plants and also not for all scenarios. In all scenarios power generation from lignite, hydropower and RES will be handled the same way. Our approach to this will be described in the following paragraphs. The specific method of power generation from nuclear energy, hard coal and natural gas for the scenarios will be described in the next chapters.

Lignite

For the development of lignite, the following assumptions were made: Until 2010 the maximum generation from lignite will be 130 TWh, afterwards the maximum generation will only amount to 70 TWh. The reason for this assumption is that a high use of lignite is not compatible with climate policy targets, when simultaneously phasing out nuclear energy.

Renewable Energies

Several studies offer a wide range of estimates about the future contribution of Renewable Energy Sources to power generation. In our model, we estimate a growth of net power generation from renewable energies of up to about 70 TWh. That equals a threefold growth to a share of power generation of about 15% compared to today in the year of 2030. Higher shares are possible and ecologically desirable, but would be the same for all scenarios and thus are without relevant influence on the results of this study.

Reference Case: Phase-out of Existing Nuclear Power Plants

The reference scenario describes a phase-out of existing nuclear power plants until the end of 2028. It is assumed that all 19 nuclear power plants technically can be operated over the estimated lifetime of 40 years. The operating costs of nuclear power plants are estimated so high that also from an economic point of view this seems possible. The first nuclear power plants to go off the grid would then be Obrigheim at the end of 2007 and the last one would be Neckarwestheim-2 at the end of 2028.

According to today's planning, hard coal power plants will be added only insignificantly until 2006. It is being estimated that from then on, no new power plants are going to be built, until the total available net-capacity will reduce below 90 GW.

Thus, in the model some over-capacities will be reduced. From then on, exactly the additional capacity of power plants that is necessary to maintain an available capacity of 90 GW will be built. In the years 2016 until 2030 this will require considerable additional erection of power plants in an order of about 35 GW. On one hand the closed down power plants from the 80s have to be replaced and in addition new power plants replace the part of the reduced nuclear energy capacity that will not be compensated by erecting natural gas power plants and enforcing renewable energy.

Accelerated Phase-out

Here, we distinguish two cases: In the case of the phase-out scenario 2005 we estimate that with a law concerning the phase-out coming into force on January 1st 1999, the nuclear power plants Obrigheim (starting up: 1968), Stade (1972) and Biblis A (1974) would be taken off the grid, because by then they have reached their maximum permitted operating time. (According to the coalition agreement laying down consensus-talks with the electricity sector for the period of one year, this date would have to be delayed for one year, which would not affect the results of the study very much.) The nuclear power plants Biblis B (1976), Brunsbüttel (1976) and Neckarwestheim 1 (1976) will have to go off-grid by the end of 2000, the nuclear power plants Isar 1 (1977), Unterweser (1978) and Philipsburg (1979) follow gradually. In the year 2005 nuclear power will cease to be available for power generation.

It cannot be assumed that the short-term reduction until 2005 of 20 GW capacity in the base load area can be compensated for by existing power plants. This might be possible in calculations, but in practice, this would mean that practically all statistically registrated power plants, including those for reserve, will be used, independent of their technical condition or availability of staff, and will have to run over 6000 hours per year trouble-free. Therefore, it seems to be reasonable to assume that 50% of the power plant capacity taken off-grid will immediately have to be replaced by building new power plants.

In the phase-out scenario 2019 it is assumed that an agreement will be reached that until the end of the year 2019 all nuclear power plants will be closed down. It remains possible for the operators to continue operating the power plants up to reaching the assumed lifetime of 40 years. On the grounds of this assumption the remaining capacity is rather small and will reduce continually until 2030, so the lower production possibilities make no big difference.

Choice of Energy Source and Prices

Electricity generators always will invest in a portfolio of gas and coal to limit the increase of the gas price. This we take into account by mixing gas and coal in the future power generation, accordingly, in the simulation model in relation to their relative costs. The right mixture of this portfolio might be assessed differently by different protagonists on the market. A scientific analysis of the future market behaviour of supply and demand on the gas market is nearly impossible. But it is plausible to suppose that a fast switch to natural gas will lead to turbulent reactions on the gas markets that could be reflected in considerably higher prices for power plant gas. To show these effects we calculated the scenarios with varying price scenarios.

The Meaning of Prices

When prices change, also the relative economic efficiency of different power generation possibilities changes. At a low gas price relative to other fuels, and expecting this low gas price to remain stable, the generation of power from natural gas is especially favourable. But for a slightly higher gas price (compared to the coal price) these relation changes fast in favour of coal. In the simulation model, the power plant stock is being optimised economically for every price-scenario. The optimal power plant distribution and the optimal planning of application of power plants is being determined year after year on the basis of existing power plants and their cost, as well as on the basis of considerations about the additional erection of new power plants.

Thus another price scenario automatically means a different power plant distribution and a different use of power plants as well as for new as for old plants.

Carbon Dioxide Emissions

When burning fossil fuels, the greenhouse gas carbon dioxide emerges. The scenarios differ considerably according to the amount of emissions of this greenhouse gas created. There is a considerable conflict between different targets of economic policy: Domestic generation from hard coal and lignite (domestic value added) is in conflict with the target of climate policy of CO_2 reduction, a high use of natural gas in power plants to avoid further CO_2 emissions, can be in conflict with the target of the power generation being as economic as possible.

We included environmental requirements in our simulation model as follows: Nuclear power plants are not emitting CO_2, therefore, for the reference path of a slow phase-out of nuclear power plants, a gradual reduction of CO_2 emissions from power generation can be assumed. First of all because new fossil power plants emit a lot less CO_2 per unit of generated power, due to their higher efficiency, secondly because renewable energy sources will be of greater importance in the future and third because until then the whole potential for CO_2 avoidance of nuclear power plants will still be available. That way, the electricity sector contributes to a significant extent to the abatement of CO_2 emissions (in the year 2015 a level of 220 Mt will be reached).

The above-mentioned conflict of targets shows that in the phase-out scenarios the maintenance of lignite conversion into electric energy is not compatible with a CO_2 target of this kind. Therefore, in all scenarios the conversion of lignite was fixed at a minimum amount of 70 instead of 130 TWh from the year 2010 on (for the reason of comparability this also applies to unlimited CO_2 emission). It is clear that due to the different CO_2 content of different energy sources (see tab. 3), gas will have to play a special role in power generation, if CO_2 emissions have to be reduced and at the same time nuclear energy is not available anymore.

Tab. 3: Emission Factors for CO2

CO_2 Emissions (Primary energy)	kg/kWh
Hard coal	0.335
Lignite	0.407
Gas	0.200
Oil light	0.267
Oil heavy	0.285

This means on the other hand that by restricting CO_2 emissions the demand for natural gas will increase strongly. Can this remain without effects on the gas price?

For the few gas suppliers in Germany, naturally an interesting market evolves, when restricting coal use for climate policy reasons and when phasing out nuclear energy in Germany. There being no alternative to natural gas - at least at the moment - it is obvious that a restricted use of coal induces an increase of natural gas prices. In the simulation model, we reflected this by allowing the prices to rise by 20% above the trend value, according to the demand for natural gas. The trend value remains constant in the course of time in the price level "constant prices", however, in the price level "high prices" it is increasing according to the Prognos-scenario of 1995 (PROGNOS 1995). How high the increase according to the higher demand of gas will be compared to the trend value, depends on the demand.

It can be assumed that the demand can grow up to a certain trend amount, without leading to price increases, but price increases could happen, when the demand exceeds this trend.

Results

In a short form, the figures show the results of the simulation calculations (see tab. 4 for the names of scenarios). Shown are the new power plants that have to be built up to the year of 2030, the power generation according to types of power plats and the total costs of power generation (all costs in DM of 1998) cumulated up to 2030. To make the comparison easier, the differences between the scenarios are shown in the table, respectively.

Tab. 4: Names of Scenarios

	CO_2 unlimited		CO_2 limited	
	Prices		Prices	
Phase-out	constant	increasing	constant	increasing
2005	5K	5S	5CK	5CS
2019	19K	19S	19CK	19CS
2029	29K	29S	29CK	29CS
In addition: sensitivity calculations with decreasing/ increasing electricity consumption and gas price increases related to quantities as well as different operating costs				

In all scenarios about 70 GW of power plant capacity will have to be newly erected within the next 30 years. In the case of constant prices, gas power plants will take over the main share; a small part will be matched by coal. However, when interpreting the results, the methodology has to be taken into account. Coal power plants would preferably be used for base load and gas power plants for centre and peak load.

For constant prices, the distribution of the power plant stock nearly does not differ at all, and also in the scenario with increasing prices the structure of the power plant stock is almost the same in all phase-out variations. The gas price increasing stronger than the coal price, the economic threshold tends to move more towards coal, so considerably fewer gas power plants will be erected. In the scenario with increasing prices this is the more economic solution.

Fig. 3: Electricity Production Costs until 2030

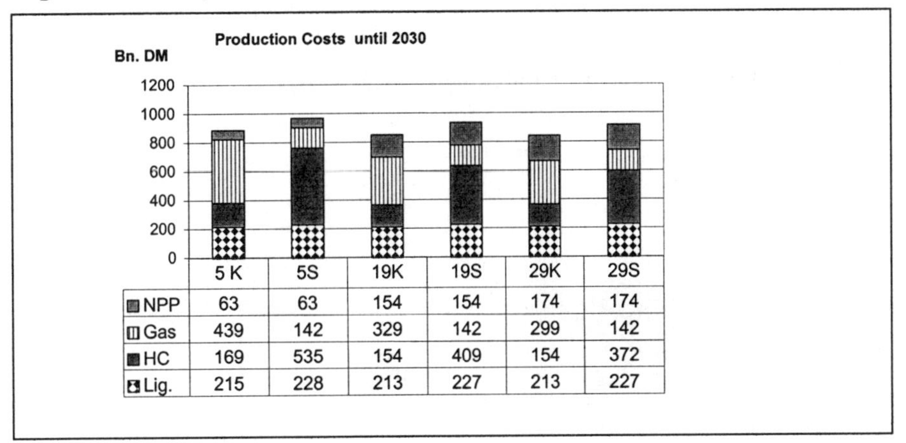

Tab. 5: Shares of Energy Sources of Production

[%]	NPP	Lignite	Hard Coal	Gas
5K	7	25	21	46
5S	7	25	58	9
19K	24	25	19	32
19S	24	25	42	9
29K	27	25	19	29
29S	27	25	39	9

As shown in Tab. 5, at constant prices and phase-out in 2005, about half of the power generation up to the year 2030 will be covered by coal, about 7% by nuclear energy and the rest by gas. In the same phase-out scenario for increasing prices, the share of gas drops towards about 9% in favour of coal.

The shares of fossil energy sources are lower in the scenarios with a later phase-out, respectively. For the phase-out in 2019, nuclear energy makes up for 24%, for the phase-out in 2029 it is 27% of the total generation. Thus, the share for gas is lower for these scenarios also with constant prices, respectively. Here about the same amount of gas power plants are being erected, but they are used later.

The costs of power generation differ according to the phase-out and price-variation. Fig. 4 shows the extra costs compared to the phase-out in 2029. For constant prices, they amount to about 46 (2005) or 10 (2019) billion DM, for the case of rising prices, extra costs are about 52 (2005) or 17 (2019) billion DM (in another study we calculated a value of 75 billion DM, but assumptions for the price development and the share of lignite were slightly different. This is shown as (prices R) in fig. 4).

Fig. 4: Additional Costs of Abandoning Nuclear Power

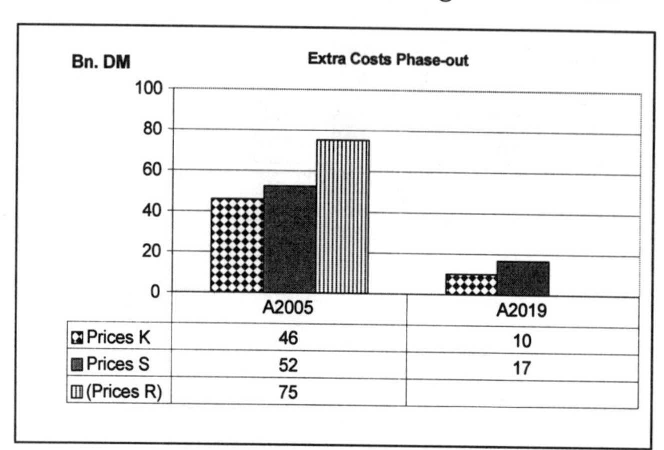

From the macro-economic point of view, the economy of Germany initially would loose these amounts of money, if abandoning nuclear energy early. As also shown in fig. 4, these amounts are considerably lower for a later phase-out, due to the reduced production potential of nuclear power plants. You have got to be aware that these costs concern the whole period up to the year 2030. Applied to the total generation in this period, of about 14,000 TWh, additional costs of between 0.4 and 0.1 Pf/kWh are the result (see tab. 6).

Tab. 6: Additional Costs in Pf./kWh

Extra Costs	Pf./kWh			
(until 2030)	5K	5S	19K	19S
CO_2 unlimited.	0.3	0.40	0.1	0.1
CO_2 limited	0.6	0.8	0.1	0.2

But these are not the whole costs that have to be considered. An early phase-out of nuclear energy leads to increasing emissions of the greenhouse gas carbon dioxide. The extra emissions depend on the time of the beginning of the phase-out, and on the price scenario. The comparability of the cases is only given, if the emissions do not differ. For the economy there are two possibilities of achieving the particular emission avoidance:

- Either additional effort in other areas of the economy is made to reduce CO_2. This leads to corresponding extra costs in those areas. STE of the Forschungszentrum Jülich recently presented calculations that have been made with the so-called IKARUS-model. The model comes to the result that short-term (until 2005) extra costs for CO_2 reductions amount to 5 to 10 billion DM per year, given the 25%-reduction target. However, this is only the case for a short-term phase-out. For a longer-term phase-out, many more possibilities for legal changes and new laws exist, which will take place anyway, for climate gas reductions, so additional costs could be much less.(MARKEWITZ 1999)
  But on the whole, the calculation presented here underestimates the extra costs, because they are being compared to the reference scenario, which itself is a nuclear energy phase-out scenario. Also, when phasing-out nuclear power plants, the reduction of CO_2 provided by them permanently has to be compensated by other means. Here, in the future additional studies sure will be necessary to put the replacement options of CHP and renewable energy sources in the framework of an economically oriented scenario.
- The second possibility of CO_2 reduction that we will present is to choose the distribution of energy sources in the phase-out scenarios so that it leads to the same CO_2 emissions in the different variations. The extra costs of CO_2 reduction then have to be borne by the electricity sector and not by other economic sectors. As described above, we calculated all variations of phase-out and prices also with such a " CO_2 break".

For the phase-out paths with a corresponding CO_2 restriction natural gas plays an even more important role than for a phase-out without such a restriction. In all variations, mainly gas power plants and only to quite a lesser extent coal power plants will be erected. The use of power plants differs slightly according to the prices of fuels, but the differences are relatively small.

Tab. 7: Shares of Energy Sources (CO_2 Restriction)

[%]	NPP	Lignite	Hard Coal	Gas
5K	7	25	13	55
5S	7	24	14	54
19K	24	25	19	32
19S	24	25	24	27
29K	27	25	19	29
29S	27	25	27	21

From the difference in the costs of phase-out paths with and without restriction of emissions of carbon dioxide, the additional costs of the restriction of CO_2 can be calculated. They are shown in fig. 5. The costs of CO_2 avoidance are about the same range as the costs of phase-out without such a restriction. So phase-out costs more or less double. However, these costs are relatively low compared to a later phase-out. That is, because here we compare with respect to the reference path, which itself is a nuclear energy phase-out path.

Fig. 5: Costs (CO_2 Restriction)

Production Costs until 2030 (Bn. DM)

	5 K	5S	19K	19S	29K	29S
NPP	63	63	154	154	174	174
Gas	439	142	329	142	299	142
HC	169	535	154	409	154	372
Lig	215	228	213	227	213	227

Fig. 6: Additional Costs Phase-out (CO_2 Restriction)

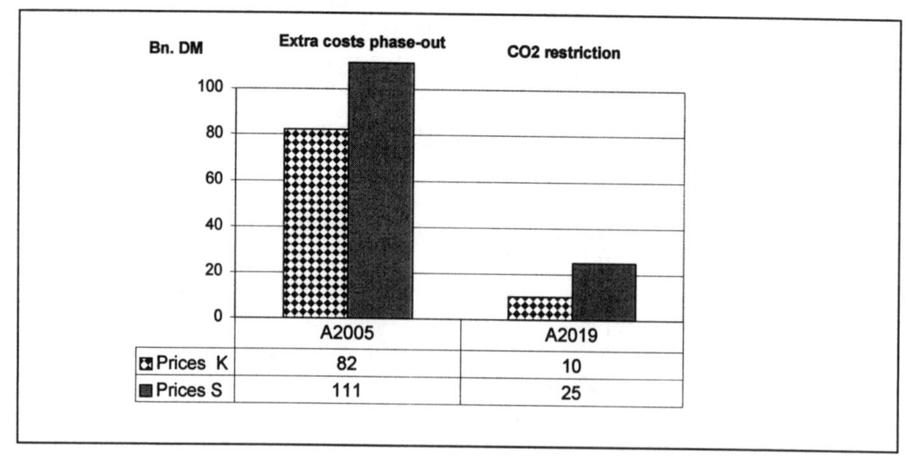

	A2005	A2019
Prices K	82	10
Prices S	111	25

Fig. 7: Costs of Avoiding CO_2

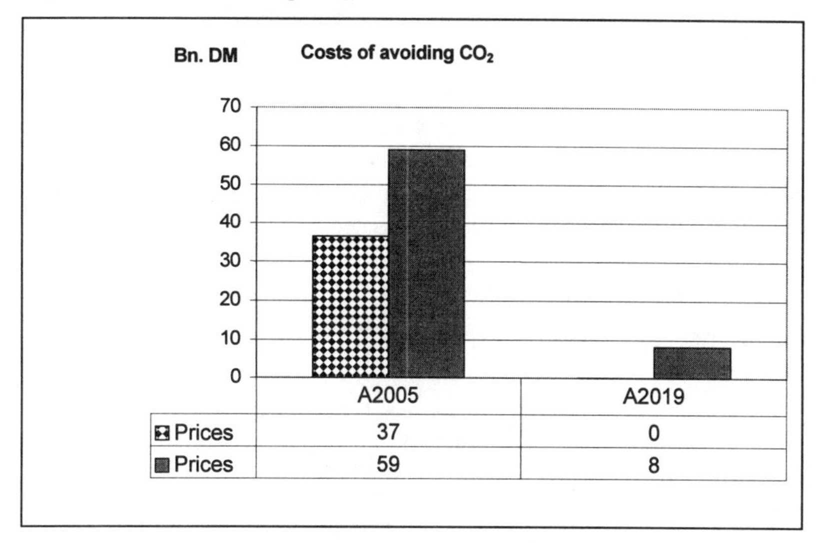

Two conclusions can be drawn from that:

1. A reduction of CO_2 emissions in the electricity sector also is possible to a certain degree with natural gas.
2. The important role natural gas will have to take over in that case, will very much increase the price risk for changes in natural gas price.

2.3. Energy Security and the External Dimension of Energy Supply

On account of the manifold interactions, there are a lot of effects when there are changes in the energy supply system. If, for example, gas power plants replace coal power plants, the following lines of influence can be directly identified: The investment requirements for a natural gas power plant of the same capacity are considerably lower, and this has effects on the power plant industry. If a new power plant replaces domestic fuel by imported fuel, there are repercussions for domestic energy production and the employment situation. On the other hand, the possible price effects induced have an effect on the economy. If electricity becomes cheaper, the electricity customers save money that can be spent for other purposes. The whole structure of consumption, and, depending on that, also the structure of production and of imports can be affected.

In the following, we describe some of the economic effects, but they do not show all possible repercussions. The analysis of employment effects is based on a macro-economic input-output-model. With this model, the employment factors, which are important for the different effects, and which have to be considered, have been determined. They tell how many employees directly and indirectly are necessary in the economy for the production of a certain value.

Balance of Payments Effects

Tab. 8 shows the additional imports of fuel necessary in the phase-out scenarios (compared to the reference case). This contribution to the balance of payments can be desirable or not depending on the particular macro-economic situation, either way, the additional payments have to be compensated by additional export profits.

Tab. 8: Balance of Payments Effect

Additional imports compared to reference case	**5K**	**5S**	**19K**	**19S**
	Billion DM			
CO_2 unlimited.	121	94	21	18
CO_2 limited	134	183	21	40

Employment Effects

Compared to the economy as a whole, the electricity sector is a very small sector. Nevertheless, big changes, as e.g. a fast change in the structure of the power plant stock, have quite a noticeable effect, because the capital intensity of the electricity sector is especially high. We analyse the macro-economic effects with an input-output-model. The changes in the structure of the power plant stock induce investments lead to orders at suppliers of investment goods, their supplier's etc. Thus they create demand.

On the other hand a change in the structure of the power plant stock, when phasing-out nuclear energy, also leads to extra costs. These mean higher expenditures for the electricity customers. For households, those expenses compete with other expenses to be paid from their budget, and would tend to reduce the latter. For companies that compete internationally, and those needing larger quantities of energy, as e.g. in certain industrial sectors, the production costs rise, so the international competitiveness suffers. On the whole, quite complex repercussions could result (PFAFFENBERGER 1995).

Fig. 8: Employment Effects

1. Investment effect
   Investments in new plants for power generation induce direct and indirect employment effects.
2. Operating effect
   Maintenance and operation incl. provision of fuels.
3. Budget effect
   Extra costs of phase-out replace other expenses of electricity customers.
4. Regional effect
   In relation to location there are changes in the form of different direct and indirect employment as well as by changes in the local tax revenue: additional employment at the locations of replacement power plants, losses at the location of nuclear power plants.
5. Dynamic effect
   Market changes caused by adjustments to a different price level and other availability with the corresponding effects on the structure of industry.
6. Foreign trade effect
   a. Export / import of plants, fuel and electricity
   b. Different currency receipts for suppliers (plants, gas, and imported coal) with consequences for their purchases of goods in the home country (international budget effect).

Some important effects have been listed in fig. 8.

The investments necessary for changing the structure of the power plant stock can be determined very exactly from the scenarios. But for economic effects the extra costs are also relevant. They are assessed in a rough calculation, assuming that extra costs only lead to increasing prices for tariff customers. As described above, increasing prices are less easy to be pushed through for electricity customers who compete internationally.

Fig. 9: Loss of Jobs

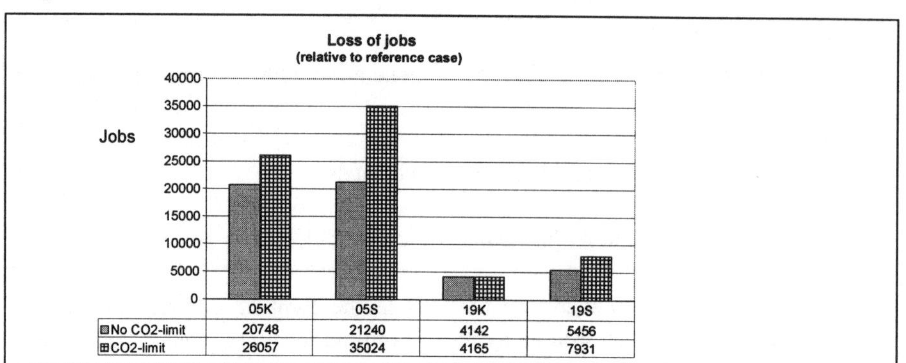

Fig. 8 points out the effects on employment in the phase-out scenarios (PFAFFENBERGER, 1998). They mainly show the calculated losses of jobs of the different phase-out scenarios. Two effects mainly can explain them:

- Extra costs of electricity have a negative effect on employment (budget effect),
- When phasing out nuclear energy, the high operating costs of nuclear power plants, which to a great extent are domestic value added, become unnecessary. The operating costs of replacement power plants (natural gas or coal) are considerably lower and amount only to 25% of the operating costs of nuclear power plants. The operating costs contain direct employment at nuclear power plants (this is only a small number of persons) and all suppliers who render maintenance and repair services.

Especially for a short-term phase-out, the corresponding losses of jobs are high, because the corresponding domestic value added is lost for a longer period of time. In the replacement power plants, mainly imported fuel is used, which only leads to a small amount of value added in Germany, caused by the necessary transport.

Looking at the whole period of time, the investment effects do not matter a lot, because investments are about the same amount in all scenarios over the total period of time. However, they emerge at different times. During the corresponding periods there can be very different effects, which in this context cannot be described in detail.

2.4. Phasing out Nuclear Energy

The objective of scenario calculations is not to make forecasts, but to identify relevant areas of problems where various alternatives are considered. Therefore, here are some concluding remarks.

1. The aim of our analysis was not the overall assessment of nuclear energy as a source of energy. For that a long-term and global analysis would be necessary. In reality the perspective of nuclear energy in Germany heavily depends on political decisions to come. Existing nuclear power plants are competitive, but building new ones is highly unlikely under competitive conditions. Only by government intervention could the nuclear option be maintained. This is true for the British model of competition as well as the German "cul de sac" in energy policy. If a society decides pro-nuclear it must follow the French, Korean or Japanese way. Private investors in nuclear power need security.
2. If this phase-out (disregarding economic aspects) is to come, the question is how to phase out the plants operated at present. These plants are able to provide contributions to power generation over a longer period of time, which in total could even be higher than the contribution they have already made.
3. A fast phase-out of nuclear energy means destroying a considerable amount of capital. The assessment of this capital depends on the assessment of the replacement of nuclear power plants. That depends e.g. on future energy prices and especially on the pace of phase-out. Here we calculated very different scenarios. A scenario of replacing nuclear power to a great extent by gas power plants offers favourable chances at today's prices and costs. At the same time such a strategy requires natural gas to be available in sharply rising amounts without changes of price. The last condition seems not very probable in view of the natural gas market structure.
4. A mixed strategy for the substitution of nuclear energy by a mixture of natural gas and coal is more acceptable considering price and cost aspects, but it raises larger problems with climate policy, because much higher emissions of greenhouse gases are related to such a strategy.
5. A phase-out neutral to climate issues raises the costs of phase-out. Because either other economic sectors have to contribute more to greenhouse gas reduction, than in case of nuclear power being available, or the electricity sector itself one has to increase the use of natural gas to be able to reduce greenhouse gases. This means refraining from creating a great part of domestic value added in the area of coal production.
6. A higher contribution of renewable energy cannot solve the problem: The use of RES is not directly related to the emission of greenhouse gases. If nuclear energy could be replaced by renewable energy sources, the climate balance

would remain neutral in the long-term. But this would not be a contribution to the reduction of greenhouse gases.

7. There are large potentials for combined heat and power generation in Germany, which are not yet exploited. In connection to decentralised plants that could provide cheap heat, such plants could make an important contribution to power and heat generation in the future. Under today's economic conditions big obstacles for the introduction of such plants exist, because the market does not remunerate the environmental advantage of co-generation.
8. When phasing out nuclear energy, energy policy faces the problem of combining environmental soundness and economic soundness in the framework of a global economy as well as the stability of political conditions. Considerable tasks of adjustment in legislation are related to this. Macro-economically seen, it would be a great advantage, if the earning power from nuclear power plants - as far this is compatible with security aspects - could be invested in these constitutional change tasks. A social pact with a prolonged time of operation for nuclear power plants in connection with an emphasised beginning of support for RES and CHP, as well as the exploitation of energy saving potentials, would be an economically rational way of approaching this.

2.5. Issues of Reprocessing Nuclear Fuel and the Treatment and Storage of Nuclear Waste

For nuclear energy use, the purchase of fuel takes place analogously to the purchase of other primary energy sources on international markets. But for nuclear energy the governmental influence and the disposal of nuclear waste is specific. Research and development (R&D) especially for basic research as well as the governmental supervision are important. The motives are protection of the population as well as control of dissemination of material that could be used for the production of arms (non-proliferation), which is included in international contracts (e.g. non-proliferation treaty, EURATOM). Both aspects also are relevant for the disposal of nuclear waste. Here, another problem is that there are several decades between power generation and final storage, depending on the technically-physically necessary intermediate storage. Besides, the necessity for disposal will arise independent of the further future of nuclear energy, because waste has been produced in the past and the reactors themselves have to be pulled down.

The nuclear fuel cycle consists of the entire process chain, from exploitation to the use in the reactor and towards disposal. The supply side (front-end) comprises extraction, conversion, enrichment and fuel element production.

The fuel demand of a reactor is technically fixed by capacity, load factor and mass burning rate (energetic use of fuel) in quantity and quality (enrichment). Nuclear fuel is being "consumed" by using it in the reactor, so annually - mainly when revising the nuclear power plant, when it is off-grid anyway - a certain share of the fuel elements (FE) will be replaced by new ones.

After discharge of burned fuel elements (SNF = spent nuclear fuel) the disposal stage of the nuclear fuel cycle (back-end) begins. According to general understanding, there is a choice between direct ultimate waste disposal or reprocessing. At ultimate waste disposal the SNF will be stored externally after it had been stored in a nuclear power plants' own intermediate store (wet storage approx. 5-7 years) and will be stored ultimately after the end of the period of intermediate storage (approx. 25-40 years) as well as a particular preparation (conditioning). At reprocessing, the spent fuel elements will be mechanically dismantled; the fuel will be dissolved and separated into used uranium, plutonium and waste. The plutonium will be supplied to the nuclear fuel cycle again, together with (natural) uranium in so-called mixed oxide fuel elements (MOX), while the used uranium can be used again after conversion, enrichment and FE-production in a special form of fuel elements.

For reprocessing, the demand for natural uranium is lower compared to ultimate waste disposal; however, a conclusion about the economy of reprocessing cannot be drawn from this statement about the alleged "protection of resources". In the context of an overall assessment, reprocessing would only be economic if the additional costs were lower than the costs for fresh uranium. It is controversial how many times uranium can be reprocessed. The tendency towards a higher mass burning rate changes the composition of the spent fuel, which makes reprocessing technically more difficult and more expensive. If today ultimate waste disposal were cheaper, economic logic would demand the choosing of ultimate waste disposal, which for the time being will be long-term intermediate storage. Later, technically it would still be possible to opt for reprocessing.

At reprocessing the volume of radioactive waste is higher than at ultimate waste disposal, but it is mainly low and medium active waste (LAW = low active waste, MAW = medium active waste). On the other hand, the spent fuel at ultimate waste disposal will be stored as high active waste (HAW). When reprocessing capacity is limited, this can be interpreted as a delayed ultimate waste disposal. Therefore the classification of reprocessing as a "closed nuclear fuel cycle" and of ultimate waste disposal as an "open nuclear fuel cycle" is misleading.

While HAW is exclusively created when using nuclear energy, MAW and LAW also evolve from other sources (e.g. medicine, research, industry, agriculture, and military) to quite an extent.

In addition to the waste from the nuclear fuel cycle, there is further waste to dispose of, from operating the nuclear power plant (operating waste, like suits, gloves etc.) and after closing down the nuclear power plants, from their dismantling. They differ in their radioactivity. The main share of radioactive waste develops only very little heat (MAW or LAW). They are ultimately stored in several countries, e.g. the United Kingdom, France, Finland, Sweden, South Africa and Germany in near-surface storage. The final waste storage of waste from reprocessing or ultimate waste storage is not yet being realised. For HAW the final storage in deep geological formations is favoured worldwide. The German waste disposal concept aims at storing this waste in the salt dome of Gorleben, which still has to be examined as to whether it is suitable, and the use of which still is controversial.

On the supply side (front-end) Germany is not very present. It is dependent on international markets for natural uranium. Within Germany there have only been a plant for enrichment operating in Gronau and a fuel element factory in Lingen since 1995. Further plants (e.g. MOX-production in Hanau) have been closed down or did not even begin operation. The former uranium mine of the GDR (soviet-german Wismut) has been closed down as well.

Plans for the construction of a commercial reprocessing plant (in Karlsruhe a small plant for research has been operated) came to nothing in 1979 (Gorleben) and 1989 (Wackersdorf). Because until 1994 reprocessing was required for operators by the nuclear energy law (Atomgesetz), contracts with foreign companies like the French Cogema (reprocessing plant in La Hague) and the British BNFL (Sellafield) were signed. These contracts are long-term and contain the option of prolongation over the millennium. Since 1994, the nuclear energy law allows the choice between reprocessing and ultimate waste storage.

The question of whether reprocessing is economic or not, can be discussed by comparison with a fuel equivalent fuel element. Reprocessed material leads to capital and energy costs, as well as wages for the extra effort, compared to ultimate waste disposal, but is a substitute for new uranium. The cheaper uranium becomes and the cheaper enrichment is by low costs of separation, the cheaper a fuel element from primary material becomes.

The margin for fuel elements from used uranium or rather MOX-fuel elements decreases, because the production costs for fuel elements are much higher, as for fuel elements from primary uranium (on account of the unfavourable mixture of radioactive and toxic material)

The German waste disposal concept at present aims at three sites of final storage. The ultimate storage Morsleben (ERAM) in Sachsen-Anhalt has been taken over from the former GDR. It is only suitable for certain kinds of waste with low development of heat and its operation is limited until the year 2000.

The former mine Schacht Konrad (Niedersachsen) is in the process of getting a licence procedure as an ultimate storage project for all kinds of waste without heat development. As an ultimate storage project for heat developing waste or rather spent fuel elements, the salt dome in Gorleben (Niedersachsen) is being examined. There is an external intermediate storage for HAW/spent fuel elements in Gorleben and Ahaus (Nordrhein-Westfalen). In addition there is an intermediate storage for LAW/MAW (e.g. Gorleben, Mitterteich) and an East-German intermediate storage in Greifswald. Waste that does not come from nuclear energy normally is stored at collecting sites in each federal state.

The existing waste disposal system is not based on a closed concept, but rather the result of historic developments. Originally, the salt dome in Gorleben was planned to be the (only) final storage for all kinds of nuclear waste. Until its starting, which was initially planned for 1988, the use of Schacht Konrad was meant to be a buffer. Facing much lower amounts of waste, as well as the uncertainties about the future of nuclear energy, already from the point of view of costs the question arises of whether the system with simultaneous operation of two sites of ultimate storage will be realised.

For the costs of waste disposal there is the "polluter pays principle". For future costs (including dismantling) operators are building reserves. Costs for today's projects are paid by the operators as well, according to a special regulation (Endlagervorausleistungsverordnung). Further parties responsible for waste (e.g. federal government, federal states for research plants) also contribute to the problem. Waste disposal is a considerable cost factor for nuclear energy. It is estimated that the dismantling costs for reactor amounts to about 15-20% of the construction costs. For the costs of the nuclear fuel cycle, the back-end side has a share of approx. 60% compared to 40% for supply: The "disposal of ashes therefore is more expensive than the fuel". (HENSING 1998)

2.6. Forms and Effects of an Ecological Tax Reform in the Netherlands, Denmark and Sweden

In the Netherlands energy sources are taxed with a general value added tax and additionally with four energy-oriented taxes. The general consumption tax on energy and the tax for financing oil reserves are not counted as ecological taxes due to having mainly a financing character.

However, the Environmental Tax on Fuels taxes petrol, diesel, light and heavy heating oil, natural gas, coal and uranium-235, when used as fuels, as well as certain input material in the industrial sector. Electricity is taxed only indirectly, by taxing certain input fuels for generation. In addition to the energy component also the CO_2 content is being taxed.

The tax has to be paid when exploiting, producing or importing energy sources to be used as fuels. Despite power imports being favoured by the indirect burden on domestic power generation due to taxing primary energy, there is no compensating import tax.

Since 1992 the use of the tax revenue for certain ecological expenditures is not ring-fenced, but flows into the general budget. There are no special regulations for industry concerning competition, only for large customers there are more favourable tax rates from a certain amount onwards.

Because this tax did not produce the desired effect on CO_2 reduction, a second tax, the Regulatory Tax on Energy was introduced in 1996. This tax is imposed on natural gas, petroleum product (e.g. light heating oil) liquid gas and power. The Regulatory Tax on Energy is also a combination of an energy tax with a CO_2 component. The price for power rises by about 15%. Tax is paid by companies when selling to the final consumer. There are minimum and maximum limits, to prevent on one hand negative effects of distribution as a burden for low-income groups and on the other disadvantages in competition for large customers of power and gas.

The problem is that here under certain conditions an increasing energy consumption can be an advantage. The tax is related to the pre-condition of implementing compensation neutral to the states' revenue, which is done with different measures. Large companies can sometimes achieve tax relieves higher than the tax burden.

In addition, there are exemptions for certain sectors (greenhouse operators) and special regulations for renewable energy. The tax reaches approximately 95 % of all enterprises and private households. For imports, similar taxation can be assumed. Electricity is taxed doubly, because of the indirect burden of the Regulatory Tax on Energy and additionally the Environmental Tax on Fuels, on other fuels up to four ecological taxes are imposed at the same time.

Concerning employment the ecological tax reform in the Netherlands has had a slightly positive effect, but concerning the emissions a final judgement can not yet be made. Further ecologically motivated taxes (e.g. on wastewater, waste, waste oil etc.) complement the ecological taxes. (ARNDT 1998)

In Denmark, taxes on electricity as well as diesel and heating oil were already introduced in 1977 under the impression of the first oil price crisis. These taxes were based on the energy content of the different energy sources. On heating oil there was a tax rate, which was much higher than for coal, but the tax on electricity that was divided among the consumers, was by far the highest. Partially or totally exempted from tax payments was the production of the domestic energy sources renewable energies and natural gas. On the consumption side industry and agriculture were exempted from the taxes.

In the meantime, the tax rate has been increased several times. In 1992 a comprehensive tax reform took place, in the scope of which, in addition to energy taxation, taxes on CO_2 were also introduced.

While all other areas had to pay an energy tax related to the energy source as well as the CO_2 tax, companies only had to pay the CO_2 tax. The latter could be reimbursed up to 50% depending on the share of the CO_2 burden from the value added of the company, so actually this was a far reaching exemption from energy taxes for the industrial sector. The difference between tax payments of companies and other sectors was very high, due to the system of partial or total reimbursements to companies, sometimes about 10- to 100-fold.

In 1995 this regulation was amended. The tax rate continues to be different for different energy sources. For electricity, the tax is imposed on the generated power, not on the input fuel used for electricity production; there is a combination of primary and secondary energy tax. The energy is taxed on the basis of the energy content; the CO_2 component is based on the CO_2 content and an additional sulphur tax (only on coal and heating oil) on the sulphur content.

The criteria for an energy intensive company have been redefined, now the following apply:

- The single units of a company count, not the company as a whole
- There is a difference between the production process (and here further between energy intensive and light processes) and demand for room heating
- Criterion for the definition of "energy intensive" is a share of 3% of the tax burden compared to the value added and additionally a share of 1% of turnover
- When attending an energy audit, a reduction of the tax burden is possible.

Compared to the tax of 1992 the tax rates have increased, e.g. for the energy source heating oil. An exemption is energy intensive producers that attend an energy audit. A higher tax burden is very significant in the area of room heating, where a greater incentive for energy saving has been created. The tax rates basically are increasing, but for energy intensive processes they are not dynamic.

Danish taxes are also imposed on power imports. The levy is based on the methodology for the levy of value added taxes. By decreasing the labour costs, investment incentives and other measures, the tax is meant to be neutral to the government budget. The compensation is divided between households and companies to avoid cross subsidies. The ecological as well as the economic targets do not seem to have been achieved; the desired effects on employment of the tax reform are rather low. In addition to the mentioned taxes there are other ecological taxes in Denmark such as taxes on products, waste and transport as well as for water protection. (ARNDT 1998)

Sweden is most advanced with introducing energy taxes form an ecological point of view. At the time of introduction of the ecological tax reform, other taxes have been lowered (income tax) or abolished (luxury tax). There are three areas of energy taxation:

- The real taxation of energy with a general energy tax, a CO_2 tax, a sulphur tax and a tax on electricity,
- Production taxes on hydro power plants and nuclear power plants as well as
- Emission taxes on nitrogen oxide and emissions of domestic air transport.

Taxes are imposed on primary as well as on secondary energy sources. The energy CO_2 and CO_2 tax depends on the particular energy- or rather CO_2 content and is imposed on petrol, liquid gas, heating oil, methane, natural gas, coal and petroleum coke. Petrol, diesel and heating oil are further divided into three classes, depending on the share of sulphur and hydrocarbons. Class 3 is the most environmentally harmful.

With this tax system the desired allocation effects could be achieved in Sweden. Companies invested in modern refineries, and in 1993 about 20% of diesel oil could already be classified "class 1" and 57% "class 2", while the share for class 1 and 2 together was only 1% in 1990. The sulphur tax is imposed on the same fuels mentioned for the energy CO_2 and CO_2 tax as well as on peat. It is based on the sulphur content of fuels measured in weight percentage.

Both exploitation, production and imports are taxed. There are exemptions, e.g. for certain stocks, for some biological energy production processes and also for the industrial sector, which is almost completely exempted from the tax, by possibilities of tax deduction for the energy component by 100% and for the CO_2 component by 75% – the aim is to maintain competitiveness in the exposed sector.

As far as the electricity tax is concerned, the tax is imposed on the consumption in kWh; taxes on the fuel used for power generation can be deducted, to avoid double taxing. Industry again is nearly totally exempted. All taxes also apply on imports, while exports are tax-free. Taxes are imposed on electricity generated domestically, as well as abroad; thus the domestic generators are not put at a disadvantage. The demand for taxation neutral for the budget came rather late, but it made financing of constitutional change possible.

The production taxes are imposed on hydro power plants according to the age of the plant and there is a high, strongly dynamic tax on nuclear energy. Nuclear energy is to be phased out by 2010. Here, no compensation for imports exist, so domestic operators are at disadvantage.

In addition to the two mentioned tax forms, there are taxes on emissions. Since 1992, there has been a NO_X-tax, the amount of which depends on the emissions measured during combustion. Taxes on domestic air transport depend on the distance of the flight and the type of aeroplane, but not on the fuel consumed. But, especially for nitrogen oxide, undesired processes of substitution took place towards untaxed, but environmentally more harmful, fuels were the result. In addition to the here mentioned taxes on energy there are several other ecological taxes e.g. on waste water, pesticides, an emission-related automobile tax and some others (ARNDT 1998).

3. Ecological Tax Reform: Theory, Modified Double Dividend and International Aspects

3.1. Basic Theory

Economists have discussed the options of an ecological tax reform for many years. There is indeed a broad range of literature on ecological tax reform (e.g. BOVENBERG et al., 1994, 1995; SCHOLZ, 1997; KLEPPER/ SCHOLZ/ PEFFEKOVEN/ VON WEIZSÄCKER, 1998; WELFENS, 1999) which suggests that raising energy taxes could internalize negative external effects from production and energy-intensive consumption while reducing labour costs from the additional tax revenue in order to stimulate more employment. As regards the objective of internalizing negative external effects in energy generation charges should be differentiated according to the specific CO_2 content (we disregard other emissions) which is rather high in the case of coal but low for gas and zero for hydroelectricity.

The theoretical and empirical debate has not always been conclusive, and there is the open question, whether or not changing bargaining behaviour under the new tax regime would reduce any positive employment effect. A critical question also concerns the potential fall of output as a consequence of rising energy prices in the context of an ecological tax reform. Since one typically assumes that industrial energy demand has low price elasticity, there should be only minor negative welfare effects from rising energy prices in the context of an energy tax reform. Furthermore there is no clear argument which ecological tax revenue should be used to cut payroll taxes. However, a general theoretically convincing argument could be that the ecological tax reform is part of a broader reform of the tax system aimed at internalizing negative external effects (requiring adequate subsidization) on the one hand while reducing highly distortionary effects on the other hand. The view developed here will be applied within our concept of an innovation-augmented ecological tax reform. CO_2 energy taxes internalize negative external effects; R&D promotion internalizes positive external innovation effects; and cutting highly distortionary payroll taxes reduces factor market inefficiencies – the overall impact of our approach should be efficiency gains in allocation and hence positive income and welfare effects. In a rigorous theoretical perspective the optimal tax reform would be characterized by equalization of negative social marginal effects from energy emissions (and labour market distortions) with the positive social marginal benefit from innovation.

CO_2 emissions are a crucial problem for the global climate as are FCKWs (LOSKE, 1996); technical change, discounting and the need to limit the speed of global warming are important aspects of modelling (AMANO, 1998). For a sustainable development worldwide it is necessary to cut CO_2 emissions in OECD countries because the natural growth and catching-up process in developing countries will bring about a long term increase of gaseous emissions anyway. As the experience of the OPEC price shock has shown – with prices going up strongly and specific energy use falling gradually as the consequence of induced innovations and substitution effects – the price mechanism can be quite useful for raising energy efficiency. This experience is an essential background for the discussion about a green budget reform in OECD countries (GALE/BARG/GILLIES, 1995; VAN DEN BERGH and VAN DER STRAATEN, 1994).

3.2. Crucial Issues

Experiences in several EU countries point to the positive impact of an energy/ CO_2 tax on the environment (SCHLEGELMILCH, 1999). Sweden, Finland, Austria and Denmark are among the countries with energy/ CO_2 taxes; typically ecological tax policies are designed in a way which impose a relatively low burden on industry as politicians are afraid of negative tax reform effects on competitiveness.

A double dividend from a revenue-neutral ecological tax reform could be achieved under certain circumstances, but theoretical analysis and empirical finding point to uncertainties with respect to the likelihood of a double dividend. The first dividend should occur from the improvement of environmental quality as higher energy/ CO_2 taxes – with charges corresponding to the negative external effects from pollution - should reduce emissions. The environmental improvement effect will be rather modest in the short term, as short-term price elasticities typically are low in absolute terms. By implication this means that tax revenue from an ecological tax is fairly stable in the short term. In the long run price elasticities are higher, and firms will react to the rise of energy prices with product and process innovations. The share of energy-intensive products in total output is likely to fall unless there is a strong offsetting effect from high-income elasticities for that product group.

The second dividend should come from the reduction of distortionary taxes, i.e. mainly the fall of social security contribution rates (more generally: payroll taxes) and non-wage labour costs, respectively. Payroll taxes are known to have distortionary effects on labour supply and the efficiency of resource allocation. This means that a revenue-neutral ecological tax reform which takes the revenue from eco-taxes to reduce labour taxes – or income taxes – should create positive output effects in the context of tax rate reductions. In an economy with high unemployment there could be an additional benefit, namely that the fall of relative labour costs contributes to higher employment. With lower taxes on labour and higher emission taxes, which are corrective, one might expect that there is always a positive welfare effect. However, as emphasized by OATES (1995) there is a typical second best problem, namely that introducing a green tax could exacerbate existing tax distortions. This would mean a negative (income and) consumption effect – i.e. a negative welfare effect. Only in the hypothetical case (SANDMO, 1975) that revenue from environmental taxes – set efficiently in a way that the tax rate is equal to the marginal social damage from pollution – would fully finance the government's budget would one not have to worry about the second-best problem mentioned above.

One can modify Sandmo's argument slightly by arguing that any tax mix, which has small distortions, is likely to go along with a double dividend for a revenue-neutral ecological tax reform. From this perspective the likelihood of a double dividend depends on initial distortions related to the structure of the tax system and the size of tax rates. We may add that the positive welfare effects of reduced emissions could be underestimated if one takes into account only the direct effect of higher energy taxes; there could be indirect effects both with respect to the health care system whereby there are lower costs in a cleaner environment; and there could be positive wealth effects to the extent that reduced emissions typically go along with an increase in the value of real estate. For welfare analysis it matters, of course, whether wealth directly enters the utility function – similarly, whether the risk of unemployment enters the individual utility function. However, a revenue enhancing tax reform would be adequate (Pareto-superior) if – given politically fixed [other] expenditures – outlays for R&D support would be below the level at which the cost of innovation is equal to the marginal social benefit from innovation.

Empirical analysis can shed further light on the issue of a double dividend although many studies strictly focus on income, price and emission changes. In Sweden green taxes were introduced in 1991, and a Swedish Green Tax Commission was created in 1995. Based on the Commission's general equilibrium model BRÄNNLUND (1999) gives an analysis of the double dividend issue for Sweden. Initially disregarding emission effects he found a negative welfare effect of an ecological tax reform. The equivalent variation – as an indicator of the welfare effect (indicating how much households would be willing to pay for the tax reform to occur or not) – was negative; i.e. households would have been willing to pay a certain amount if government would not impose the tax reform. Interestingly, the equivalent variation was more negative in the case of a lump sum transfer than in the case of reducing payroll taxes, which can be interpreted as a weak case of the double dividend hypothesis. One should note that general equilibrium models are difficult to apply to conditions of high unemployment, as the implicit market clearing assumptions of the model might be inconsistent with the problem at hand.

3.3. Innovation-Augmented Tax Reform

From a theoretical perspective positive or neutral growth effects of tax reform are more likely if part of ecological tax revenues are used to promote higher R&D expenditures as was suggested by WELFENS (1999b). With higher R&D

expenditures the knowledge capital stock of society will increase and output growth will be reinforced; i.e. the negative output impact of the rise of energy prices will be mitigated or even fully offset. In the simulation approach we will not focus on the issue of an optimal tax split; rather we will impose the requirement that higher energy taxes should not reduce output relative to the base scenario (without new energy taxes).

The modelling approach of the innovation-oriented ecological tax reform can be summarized by the following elements (see figures):

- The goods supply side can be represented by an aggregate supply function where real output Y depends on the stock of real capital K (machinery and equipment, buildings), energy inputs E, the R&D capital stock R (reflecting accumulated knowledge in the context of R&D expenditures), labour L and the natural capital stock N; the latter could be a scale variable in the context of an extended simulation model. The more energy is used the higher are certain emissions – above all CO_2.
- Ecological tax reform: This tax reform is associated with a tax on the use of energy or a charge on CO_2 emissions where the revenue is used by government to replace part of social security contributions of firms; i.e. non-wage labour costs are reduced which stimulates labour demand (there will also be a positive supply effect). The implied employment growth – going along with reduced demand for energy inputs – stimulates aggregate demand; however, as cheaper labour partly substitutes for real capital, investment demand will fall so that aggregate demand might indeed be reduced. Note that this could mean a new goods market equilibrium with a lower output level which is consistent with reduced energy and capital inputs.
- Innovation-enhancing government expenditures: In the context of an innovation-oriented ecological tax reform part of government revenue is used to finance higher R&D expenditures, which should help to internalize positive external effects of R&D. It is well known that the marginal social benefits of innovations are higher than the private benefits appropriated by the respective innovator. We assume that initially the optimal R&D shock has not yet been reached. If R&D expenditures are increased – relative to GDP – there is possibly a negative employment effect because a higher R&D capital stock will raise labour productivity (and capital productivity) which in turn will reduce the demand for labour in a setting with a given aggregate demand (or a vector of sectoral demand in the context of an input-output model used to generate aggregate demand in a bottom-up approach). As regards the supply side of the goods market the rise of the R&D capital stock will raise potential output (ADAMS, 1990). To the extent that a higher R&D capital stock stimulates consumption – a higher R&D intensity goes along with more product varieties or with higher wealth as households, owning joint stock companies, anticipate

higher profits from intensified Schumpeterian competition – aggregate demand will increase. This also holds if a higher R&D capital stock stimulates net exports, mainly through higher exports from R&D intensive industries. In the context of a bottom-up macro model, relying on input-output analysis, it would be possible to analyze not only aggregate effects but also crucial sectoral impacts of an innovation-oriented ecological tax reform; sectoral shifts can be analyzed.

- To the extent that R&D capital accumulation leads to positive international technology spillover effects an innovation-oriented ecological tax reform will have positive growth effects on the rest of the world. From the literature (BAYOUMI/COE/HELPMAN, 1999; COE/HELPMAN, 1995) it is known that knowledge is diffusing internationally via exports of technology-intensive intermediate products, technology trade, R&D networks and capital goods (JUNGMITTAG/MEYER-KRAHMER/REGER, 1999). Since Germany is a major exporter of technology intensive products an innovation-oriented ecological tax reform lets one indeed expect significant external growth effects for EU partner countries and OECD countries, respectively. In a macro-model this international growth bonus effect could be modelled in the form of a higher exogenous world import demand – part of which will benefit the exporting sectors of the country with the innovation-oriented ecological tax reform.

Fig. 10: Building-Blocs of Innovation-oriented Ecological Tax Reform

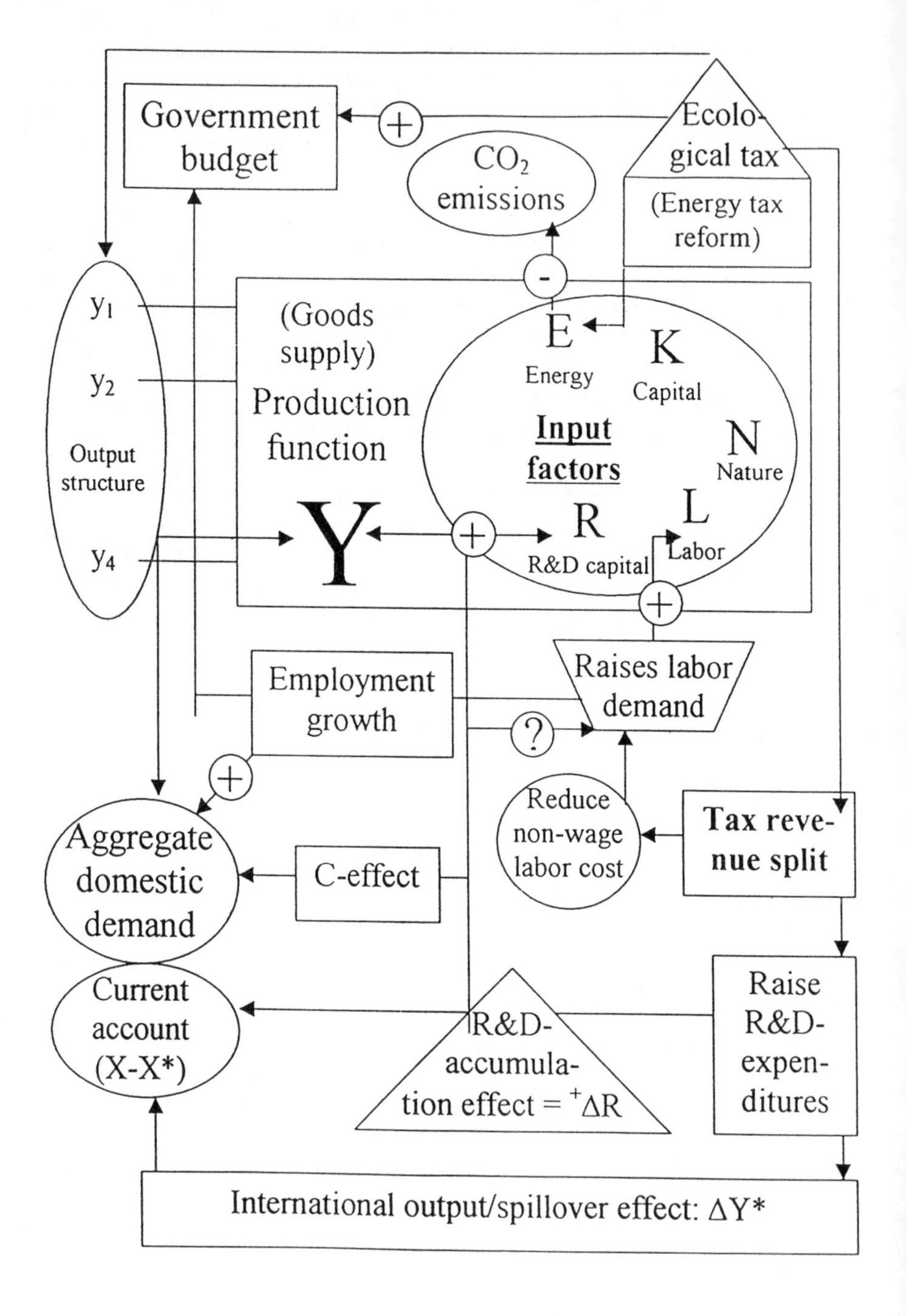

3.4. The Modified Double Dividend

3.4.1. The Traditional Approach: Reducing Labour Costs

A traditional approach for creating more employment is to cut wages and labour costs, respectively, so that more jobs become profitable. However, wages are not only a cost element but also determine aggregate demand. Therefore cutting wages is adequate only if there is a large excess supply of labour. Naturally, one would have to carefully consider differentiation across sectors, regions and skills. Moreover, the efficiency-wage hypothesis, which emphasizes that workers´ efforts are raised if relatively high wages are paid, lets one expect that a fall of labour productivity could be the result of strong wage cuts. This is not to deny that Germany and other EU countries need more wage differentiation and more labour market flexibility (ADDISON/WELFENS, 1998). However, policy must be carefully designed and should not overlook the recent analysis on the theory of employment protection (GROENEWOLD, 1999); and it should take into account that government´s R&D policy offers opportunities for raising factor productivity, including labour productivity. The equalization of real wages and marginal labour productivity cannot only be achieved by cutting wages but also by raising labour productivity.

3.4.2. The Schumpeterian Approach: Raising R&D Support

The positive impact of technological change and innovation on fostering economic growth is generally acknowledged today. Although the growth enhancing effects of new products and processes had been known for some time, it took some decades to attract the interest of researchers to study technical change. This lack of interest may be explained in part by complex procedures ruling science and technology (S&T) and the unknown mechanisms translating innovations into broad-based economic effects. However, if it is a matter of fact that technological change is a driving force behind economic growth, any analysis of macro-economic policies should take these impacts into account.

Thus, it is not surprising that recent approaches in growth theory pay much attention on technological change or its "mate": human capital respectively knowledge. The basic models of new growth theory which are in the meantime standards in modern textbooks are presented in ROMER (1986), LUCAS (1988) and ROMER (1990). A large part of new growth theory assumes a beneficial know-how "transfer" from a knowledge-generating sector, which performs R&D to the sector of the economy in which companies simply adopt it. Part of this knowledge as a result of R&D efforts is paid for by the receiving firms while some part diffuses without appropriate compensation. Thus, external effects of knowledge creation (so called spillover effects) are followed by increasing returns in production of the remaining sectors and cause all-over economic growth. One essential difference between neo-classical and new growth theory may be found in these growth-creating effects. This recent line of research regards national growth to be independent of stocks of knowledge and human capital elsewhere. Thus, economies with own knowledge- or human capital-creating sectors are growing faster in the long run than those without.

New growth theory is no exception to other economic modelling, as it does not pay much attention to the details either of what generates external effects in innovation or of the channels which link knowledge generation and adaptation (see e.g. JAFFE/TRAJTENBERG/ HENDERSON, 1992). Moreover, to switch from the inward perspective of new growth theory to a more outward "global" perspective seems to be useful, because it would be highly unrealistic (particularly for developed countries) – for the economies of the European Union, it would be simply wrong – to assume that knowledge flows will not leak out of the area delimited by national borders. In view of the increasing share of trade in world wide production and the recent surge in the exchange of production factors, technological as well as economic developments are influenced to a non-negligible degree by other economies via the world markets. In this respect think, for example, of those channels where scientific and technological knowledge accompanies exports of goods and services, the mobility of human capital within global firms or the policy of the European Commission to support preferentially trans-border R&D. An extensive discussion of the trend, motives and consequences of the globalisation of R&D and technology markets can be found in JUNGMITTAG/ MEYER-KRAHMER/ REGER (1999). Here, a view on new trade theory and some branches of evolutionary economics following the Schumpeterian tradition, which have in common a certain overlap with traditional theory, but stress the importance of technology and innovation as complementary determinants, can provide additional insights (see JUNGMITTAG/ GRUPP/ HULLMANN, 1998 and GRUPP/ JUNGMITTAG, 1999).

As far as new trade theory is concerned, a model that has been developed as part of a comprehensive analysis by GROSSMAN/ HELPMAN (1991, chapter 9) is particularly instructive. It deals with the situation most common to high technology trade among OECD countries. The focus is on the long-term growth prospects of countries opening up – step by step – to different degrees of market integration. Basically, the model is built up according to the following principles: Countries are "endowed" with labour, human capital and technological knowledge. To keep the analysis of the model's main properties simple, Grossman and Helpman restricted complexity in that the economy consists of one sector only. The focus is set on the working of integration – not on structural change within any one country. Technological knowledge generates external effects and increasing returns for the production of traded goods. In the long run, adding some further – more technical –

assumptions, growth rates depends on innovation rates – that is, on the speed with which new technological knowledge is built up.

Integrating two economies similar (or even identical) in terms of traditional endowments would lead to either unchanged trade patterns and growth rates or to increased specialisation and higher growth rates in both countries. The dynamic properties of this model heavily depend on the characteristics of the stock of accumulated knowledge before integration. Because of similar endowments with traditional factors the only difference before globalisation lies in the degree of knowledge specialisation in different areas. Given hat both economies are completely specialised on complementary fields of knowledge, integration will have no effects, neither on technological, production and trade pattern nor on long-run growth. Instead, if the stocks of knowledge have a certain overlap in both economies (e.g. knowledge accumulated in the same fields of science and technology) integration will weed out these "inefficiencies". Each country specialises on one part of this knowledge available to both economies via full integration of markets. In this situation growth is higher in both countries compared with those in closed economies.

Apart from new growth and new trade theory, evolutionary economics in a Schumpeterian tradition is concerned with the relationship between technology, trade and growth. Although it lacks a consistent body of formal modelling tools, evolutionary economics has provided a lot of interesting insights into the details of the working of economic systems. Evolutionary thinking is fundamentally based on the variation-selection principle which allows to look at the dynamic properties of systems and, thus, on economic development. Basically, evolution is thought of as being generated by creating a variety of different products and processes. Selection processes (e.g. markets) then work on reducing this variety to a certain number of viable products. The diversity of evolutionary theorising cannot be dealt with here (on this see DOSI/ PAVITT/ SOETE, 1990; WITT, 1993 or HODGSON, 1993). One of the main forces that generate new products or processes (and, thereby, increase variety) is innovation and technical change.

Concentrating first on variation, empirical studies have found that higher rates of innovation lead to higher rates of economic growth (e.g. FAGERBERG, 1988). The larger the number of different products and the higher the rate of new product generation the higher the rate of long-run growth. Saviotti has worked out a conceptual and semi-formal tool to show that we are observing a constantly increasing number of different products. Higher degrees of product variety cause higher consumer utility. This is a main reason for economic growth (SAVIOTTI, 1991). This mechanism mainly works through better adaptation to specific consumer needs (higher utility) as well as through higher efficiency of production processes. When we turn to the selection environment, most studies have found tighter selection mechanisms to favour higher growth. From a theoretical point of view, tighter selection does not necessarily prove more efficient because in this case a large number of product variants, which have incurred development costs, are selected out. But this waste of resources may be compensated by long-run efficiency of fewer but superior products (see, e.g. COHENDET/ LLERENA/ SORGE, 1992, for a discussion of this fundamental problem in evolutionary economics). Market competition as one of several possible selection environments in ideal sense weeds out all inefficient types of products in order to ensure the survival of the best-fitting alternatives. Then, in face of selection, generation of new products adapts to the characteristics of the successful variants. Therefore, it is essential for economic agents to learn quickly from the fate of successful as well as unsuccessful products on the markets and, then, to develop better variants which sell at higher prices or larger quantities. Thus, the particular strength of companies comes from learning adaptation. However, learning and adaptation are fundamentally path-dependent processes. That means, the probability to learn something useful will be much higher in areas where knowledge has already been accumulated in former times. This path-dependency of technological change and learning may be observed at the level of single companies, industries, regions and countries. It does not only explain a great deal of innovation but also the dynamics of division of labour and economic development. DOSI (1982) used this basic principle for a "theory" of technological chance. Scientific and technological change is following "trajectories" until a "breakpoint" (radical change) disrupts the smooth and gradual development.

The stock of accumulated knowledge does not only consist of scientific or otherwise codified and easily accessible findings but also of acquired "tacit" practical skills. Knowledge therefore has a "public" and a "private" part. Apart from a few really globalized and highly science-based technologies the main part of worldwide knowledge has a local character in that its geographical diffusion is limited in scope because of mobility barriers to human capital or skilled labour. Accordingly, empirical studies have found a lot of evidence that the ability to learn and to innovate greatly differs between sectors, regions and countries. See e.g. PAVITT (1984), PAVITT ET AL. (1987), DOSI/ PAVITT/ SOETE (1990) and GEHRKE/ GRUPP (1994). Thus, stocks of technological knowledge differ in scope and character between economic entities over long periods of time. They can account for innovation and growth rate differentials.

3.4.3. Empirical Links Between R&D, Output, Exports and Employment: Germany, France and Sweden

R&D and Output

The empirical investigation of the effects of technological change or more generally innovations on economic growth has produced a voluminous and diverse literature. Roughly, there are three types of studies: historical case studies, analyses of invention counts and patent statistics, and econometric studies relating output or productivity to R&D or similar variables (GRILICHES, 1995). Here, we will confine ourselves on econometric studies, which used some indicator variables to approximate the impact of technological change and innovations.

First, one important input factor for technological change and innovation can serve as a proxy variable: R&D. Most research in this vein uses an augmented Cobb-Douglas production function which includes some kind of a R&D stock beside the usual production factors. The coefficient belonging to this R&D stock can then be interpreted as production or output elasticity of R&D. Alternatively, this kind of production function is transformed into growth rates and the R&D intensity (R&D/Y) is included. The parameter belonging to this R&D intensity yields the rate of return to knowledge. Similar to these approaches is another procedure where total factor productivity is calculated first. Then again, either the logs of levels of total factor productivity are linked to some kind of log R&D stock or the first differences of log total factor productivity are regressed on the R&D intensity. The interpretation of the estimated coefficients is the same as before: the regression of the levels of log total factor productivity on a log R&D stock yields a measure of the elasticity of output to knowledge, while the regression of total factor productivity growth yields a measure of the social gross (excess) rate of return to knowledge (GRILICHES/ LICHTENBERG, 1984 and GRILICHES, 1995).

A general problem for the measurement of the effects of R&D on output is that a number of externalities arise in the innovation process. Summarizing the relevant literature on this topic, CAMERON (1998) distinguishes between four kinds of externalities. First, a *standing on shoulders effect* which reduces the costs of rival firms because of knowledge leaks, imperfect patenting, and movement of skilled labour to other firms. In a wider sense international technological spillovers due to foreign trade can also be considered as standing on shoulders effect. Secondly, there exists a *surplus appropriability problem* because even if there are no technological spillovers, the innovator does not appropriate all the social gains from his innovation unless he can price discriminate perfectly to rival firms and/or to downstream users. Thirdly, new ideas make old production processes and products obsolescent: the so-called *creative destruction effect*. Fourthly, congestion or network externalities occur when the payoffs to the adoption of innovations are substitutes or complements. This is sometimes called the *stepping on toe effect*. The adequate consideration of these effects in empirical investigations offers a wide field for further research. Ups to now, these effects are only taken rather roughly and partially into account in most empirical studies.

Generally, studies which are based on time series data on levels of output and R&D stocks for individual US, French and Japanese companies found output elasticities lying between 0.06 and 0.1 (GRILICHES, 1995). Considering results of this kind of studies for Germany and France at different levels of aggregation, the estimated output elasticities turned out somewhat higher. For the total economy of West Germany PATEL/ SOETE (1988) estimated 0.21 as the output elasticity of R&D. However, in a recent study BÖNTE (1998) estimated only output elasticities between 0.03 and 0.04 for the R&D stock of selected sectors of West German manufacturing. At firm level CUNEO/ MAIRESSE (1984) estimated for the R&D stock output elasticities between 0.22 and 0.33 for France, MAIRESSE/ CUNEO (1985) estimated values between 0.09 and 0.26 and MAIRESSE/ HALL (1996) between 0.00 and 0.17. At the level of the total economy PATEL/ SOETE (1988) estimated a value of 0.13 for the output elasticity of R&D for France. COE/ MOGHADAM (1993) estimated with their preferred specification an output elasticity of 0.17 for the R&D stock of France.

When growth rates are used as dependent variable and R&D intensities as independent variables, the estimated rate of return lies – summarising the bulk of empirical results for different countries and different levels of aggregation – mainly between 0.2 and 0.5, with most of the recent estimates falling in the lower part of this range (GRILICHES, 1995). However, the results for West Germany are a little bit puzzling. At a firm level, BARDY (1974) estimated direct rates of return to R&D between 0.92 and 0.97. But at an industry level MÖHNEN/ NADIRI/ PRUCHA (1986) estimated a direct rate of return to R&D of 0.13 and O'MAHONY/ WAGNER (1996) found at the same level a direct rate of return of 0.00. With a different approach BÖNTE (1998) calculated net rates of return for selected sectors of West German manufacturing between 0.23 and 0.3. This is quite in accordance with the general results and with the results for France at the firm level where GRILICHES/ MAIRESSE (1983) estimated a rate of return to R&D of 0.31 and HALL/ MAIRESSE (1995) found values between 0.22 and 0.34.

However, most of the studies considered here simply treats R&D as another form of investment and does not allow for the effects of the externalities mentioned above. Therefore, it is unclear whether such studies under-estimate or over-estimate the effects of R&D. JONES/WILLIAMS (1997) derived an endogenous growth, which takes these externalities into account, and calibrated it to a range of plausible parameter values. They find that in most cases the excess returns to R&D (calculated as the social return minus the private return) are positive, but less than 20 per cent. If the large degree of risk and uncertainty in the innovation process as well as information asymmetries between capital markets and R&D spenders are taken into account, it is not surprising that large social returns to R&D can coincide with relatively low rates of R&D investment. JONES/WILLIAMS (1997) conclude for the USA that the optimal amount of R&D investment is about four times the amount actually invested. However, other studies found less overwhelming empirical evidences. Bartelsman et al. applied the Jones/Williams model to Dutch manufacturing firm-level data and found that the private rate of return probably under-estimates social returns by only a few percentage points. Examining the effects of R&D on productivity in a panel of French and US manufacturing firms, Mairesse/Hall (1996) found that R&D earned a normal private rate of return in the USA during the 1980s. For selected sectors of German manufacturing, BÖNTE (1998) concluded that his results provide no evidence for "over-normal" rates of returns" due to intra-industrial spillovers.

Another important source for externalities is international R&D spillovers, i.e. the impact of foreign R&D on domestic productivity and output. COE/HELPMAN (1995) captured these effects by augmenting the above mentioned total factor productivity equation with import-weighted foreign R&D stocks. For West Germany they calculated elasticities of total factor productivity with respect to foreign R&D of 0.056 (1971), 0.072 (1980) and 0.077 (1990). The elasiticities for France were a little bit lower: 0.045 (1971), 0.061 (1980) and 0.067 (1990), whereas the elasticities for Sweden were higher: 0.067 (1971), 0.087 (1980) and 0.093 (1990). BAYOUMI/COE/HELPMAN (1999) applied the same approach to a larger sample of countries and found important differences between the values of the coefficients for domestic and import-weighted foreign R&D for different groups of countries. Comparing the G-7 countries and small industrial countries, the coefficient of import-weighted R&D stocks has the same value but the coefficient for domestic R&D turned out to be much smaller for small industrial countries. For developing countries they assume that R&D capital is constant and the coefficient of import-weighted foreign R&D turned out to be clearly higher.

Up to now we do not differentiated between the possibilities of financing R&D. R&D can either be financed by companies or by government and there is a lively controversy on the effects of government financed R&D on output and productivity. In his summarising overview GRILICHES (1995) concluded that most elasticity estimates are not sensitive to whether one uses total or only company-financed R&D stocks, but that there are other indications in the data that government-financed R&D produces less benefit than privately-financed R&D. Concretely, he presents estimations where the privately versus government-financed R&D mix variable has a significant positive coefficient, indicating a the premium on government-financed R&D is smaller, but still quite large. GRILICHES/ LICHTENBERG (1984) found those spillovers between academic research and some types of government R&D and the private sector exist, but they are smaller than those between firms themselves are. ACS/ AUDRETSCH/ FELDMAN (1994) concluded that small firms (particularly high-tech start-ups) might benefit more from such spillovers. Furthermore, ADAMS (1990) found the output of the academic science base is a major contributor to productivity growth, but the time lag is approximately twenty years.

Connected to the controversy about privately or government-financed R&D are other empirical findings concerning basic research. GRILICHES (1995) presents estimation results where the basic research coefficient is highly significant and shows a rather large size. He concluded that firms which spend a larger fraction of their R&D to basic research are more productive, have a higher level of output relative to the other measured inputs, including R&D capital, and that this effect is relatively constant over time. Based on other estimation results and additional computations he concluded that the premium for basis research over the rest of R&D is 3 to 1 as far as its impact on productivity growth is concerned.

Secondly, one output of the innovation process can be used as a proxy variable for technological change and innovation: patent applications respectively the stock of patents. Such a proceeding has several advantages. On the one hand, this indicator variable avoids a lot of data-technical problems, e.g. unlike to R&D stock measures, no artificial depreciation rate must be assumed, on the other hand, it includes also the results of other knowledge sources apart from explicit R&D activities. BUDD/HOBBIS (1989) estimated for UK manufacturing long-term output elasticities with respect to a constructed stock of patents between 0.21 and 0.23. In a second paper, they estimated with a slightly different approach long-term elasticities of patenting of 0.114 for France, Germany and the United Kingdom, whereas the elasticity for Japan was 0.135 (BUDD/HOBBIS, 1989a). For the West-German business sector in the period from 1960 to 1990, JUNGMITTAG/WELFENS (1996) estimated an output elasticity of the real patent stock of 0.23. For a longer time period from 1960 to 1996 and with a slightly different approach, JUNGMITTAG/BLIND/GRUPP (1999) found a output elasticities of the patent stock lying between 0.16 and 0.19. Altogether, these results suggest that the estimates of the output elasticities of the R&D stock and the patent stock are in most cases very similar and that they contribute substantially to economic growth.

R&D and Exports

It is well known from a bulk of empirical studies that a strong link between R&D intensity and trade performance exists (see e.g. ENGELBRECHT, 1998; WAKELIN, 1998 and GRUPP/ JUNGMITTAG, 1999). However, since trade statistics are in most cases only available in a product group classification and R&D data are collected on industry level, strong concordance and aggregation problems occur in such studies. These problems are solved only very rudimentary up to now. But with a concordance between the International Patent Classification (IPC) and

SITC III it is possible to avoid this problem in an elegant manner. Such a concordance elaborated at the Fraunhofer Institute for Systems and Innovation Research contains 42 R&D-intensive (R&D intensity above 3.5 %) three-digit SITC III product groups (this High-tech list has proved to be very useful in the annual reports on Germany's technological performance for the Federal Ministry for Education, Science and Research as well as in other empirical investigations). Based on these 42 R&D-intensive product groups, table 9 shows the correlations between revealed comparative advantages (RCA) in foreign trade in 1995 and the relative patent shares (RPS) from 1993 to 1995 for the nine strong patent applicants. For large countries, the correlations between the RPSs and the RCAs are significant on the usual levels. In the cases of the USA, Japan, Germany, the United Kingdom, France, Sweden and Italy these links are significant on levels between 1.8 % for the United Kingdom and smaller than 0.1 % for Germany and Sweden.

Tab. 9: Correlation between RPS and RCA for Nine Countries

Country	Coefficient	t-value	Significance[1)]
USA	0.481	2.545	*
Japan	0.548	2.918	**
Germany	0.706	5.237	***
United Kingdom	0.415	2.478	*
France	0.366	3.537	**
Canada	0.147	0.877	
Sweden	0.547	4.424	***
Italy	0.641	3.675	**
Netherlands	0.279	0.207	

1) Significance levels: * < 5 %, ** < 1 %, *** < 0.1 % (White's heteroscedasticity-consistent estimators of the variance matrix of the regression coefficients are used to calculate t statistics).

Source: JUNGMITTAG ET AL. (1998)

R&D and Employment

A wide range of different opinions exists on the relationship between technological change or innovation and employment. Roughly examined, the discussants are allocated on two camps. One group emphasises the displacement effects of technological change. In this view, technological change increases the rationalisation potential, so productivity grows faster than production and more workers are displaced than can be absorbed elsewhere. The other group stresses the compensation effects. MEYER-KRAHMER (1992) summarises four compensation effects:

- Technological change leads to new product markets and new opportunities of employment which increase final demand and therefore stimulate employment, although this effect is primarily a result of complementary product innovations and not of substitutive ones.
- More efficient production methods reduces costs and prices (which implies raising real incomes) and increases profits, thus stimulating the effective overall demand and so compensating or over-compensating for any initial demand declines due to elimination of jobs.
- Labour-saving machines must themselves first be produced, so that the displacement of labour in the operations that are rationalising generates positive employment effects for the machine manufacturers and their suppliers.
- As a result of technological change, the ability of firms to compete internationally is improved, which has also positive effects on employment.

Which effects – displacement or compensation – can only be accessed empirically. Thereby, direct and indirect effects must be taken into consideration because labour-saving and labour-generating effects do not always coincide in time or space or do not occur in the same sector. Therefore, it is essential to examine income circulation relationships, the characteristics of specific sectors, relationships between different sectors, and the emergence and spread of new technologies over time (MEYER-KRAHMER, 1992).

MEYER-KRAHMER (1992) found in his empirical study for West Germany, that above all the R&D expenditures of the corresponding sectors have a stimulating effect on labour demand. The only exception is here mechanical engineering. Thus, increasing R&D expenditures lead first and foremost to additional employment in the original sectors. Rationalisation effects are conversely caused primarily by the purchase of investment goods with greater R&D contents. Such effects have greater importance in the automotive and textile industry, in electrical engineering and electronics, and for energy supply, communications and transportation. By contrast, these effects have only little importance for the service sector, trade and commerce. Thus, MEYER-KRAHMER (1992) concluded that technological change also leads to a shift in the structure of employment away from manufacturing towards the service sector.

These results are generally in accordance with the development of employment in the manufacturing industry in OECD countries. Figure 11 displays the long-term development of employment in total manufacturing and differentiated for R&D-intensive and non-R&D-intensive industries in Germany, France and Sweden. The data are taken from the OECD STAN database, which allows only a rather rough assessment of R&D-intensive industries. We assume that industries with a large part of R&D-intensive production are manufacturers of chemical products, non-electrical machinery, and electrical machinery and transport equipment. For all three countries under consideration we can observe similar but not identical patterns of development of employment. Generally, the R&D-intensive sectors show less reduction of employment than the non-R&D-intensive sectors. For the R&D-intensive sectors in Germany, there were only small decreases of employment after the first and second oil price crisis, which had only transitory character. In the mid 1980s a rather strong increase of employment in the R&D-intensive sectors took place with ended 1991. Thereafter, employment reduces to 5 % below its initial level in 1970. Employment in the non-R&D-intensive sectors decreased strongly and permanently after the first and second oil price crisis. A moderate increase from 1989 to 1991 was only transitory and from 1992 to 1994 we can observe a further decrease to 2/3 of the initial level in 1970. For France, there was a strong increase of employment in R&D-intensive sectors from 1970 to 1974, but only a moderate one in the non-R&D intensive sectors. Thereafter, employment decreased more or less continuously, but the R&D-intensive sectors reached 84 % of their initial level, whereas the non-R&D-intensive sectors fall on 67 % of their initial level in 1970.

The employment in the Swedish R&D-intensive sectors oscillated between 95 % and 110 % of its initial level until 1991. Thereafter it falls rather drastically to 79 %. Employment in its non-R&D-intensive sectors moved in steps with a small transitory increase in the second half of the eighties down to 64 % of its initial level.

Fig. 11: Employment in Manufacturing Industries (1970 = 100)

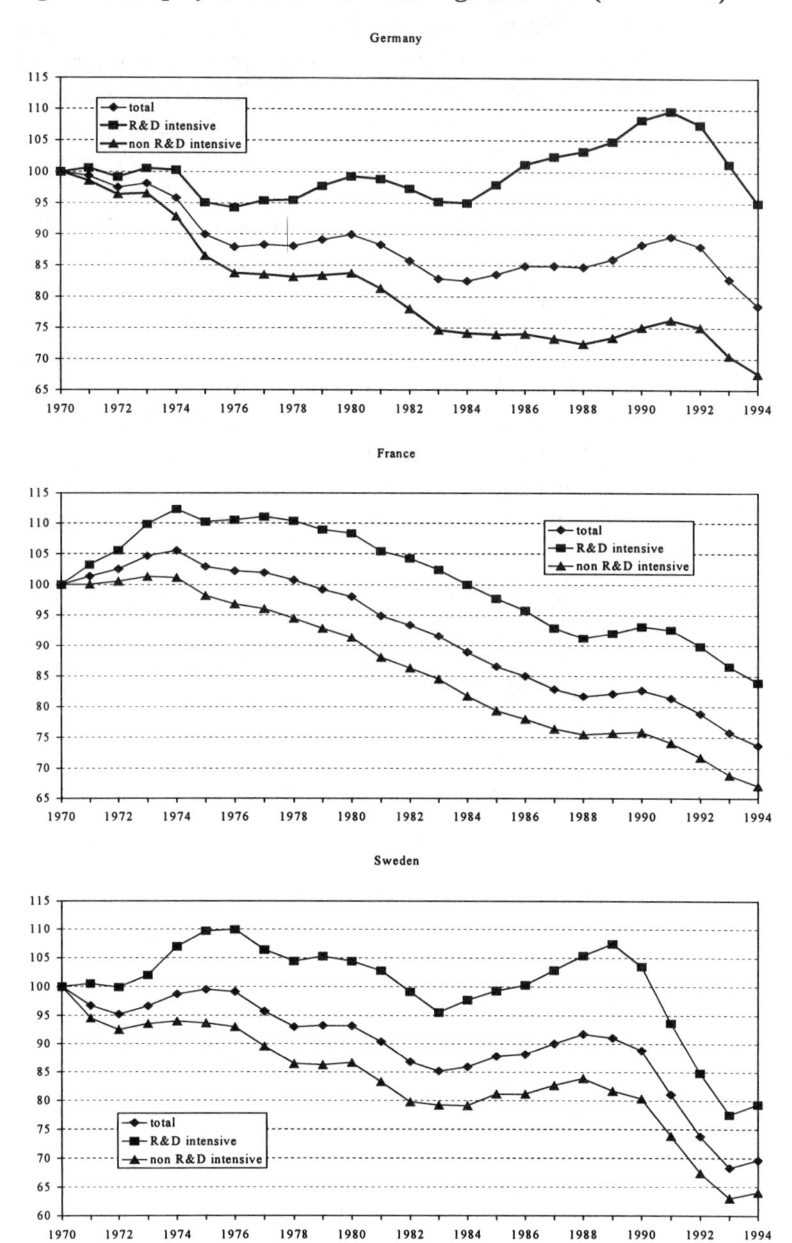

3.5. Reforms of the ESI in England and Wales – Lessons for Other European Countries

In 1989/90 the Electricity Supply Industry (ESI) in England and Wales was restructured and privatised, but the reforms under consideration was rather a process that these actions commenced, than a one-off event. Privatisation will be complete only after the BNFL is sold, which is apparently planned for 1999/2000. The ESI in other parts of the United Kingdom, *i.e.* in Scotland and Northern Ireland is structured in a different way and in part operates according to different principles. BNFL - British Nuclear Fuels Limited - is still a publicly owned company, comprising reprocessing facilities as well as Magnox Electric - nuclear power stations with Magnox reactors, a left-over from the 1996 privatisation of British Energy. The industry is subject to various regulatory interventions and operation of market forces, which to a larger or smaller extent change its structure. This in turn was one of the factors which in 1999 necessitated changes in the regulatory structure: the Office of Electricity Regulation (Offer) was merged with the Office of Gas Supply and the new energy regulator, Callum Macarthy, appointed in the Autumn of 1998 initially to regulate electricity and gas industries separately, now has the Office of Gas and Electricity Markets (Ofgem) to assist him in his duties. If one adds to this the planned reform of the electricity pool, it will soon become obvious that the years 1989 - 1999 witnessed a dynamic, evolutionary process, in which various parties - the government, the regulator, the antitrust authorities, and the firms, both from the industry and from outside it, both British and foreign, as well as the consumers - have all been using their prerogatives and freedoms, but which in its totality is no longer controllable by any single party. As a complete and thorough analysis of all aspects of this process would go far beyond the scope of this brief paper, we shall focus in it only on some issues: the initial restructuring, regulation for competition, privatisation, and the fossil fuel levy (nuclear power stations and renewable energy sources). The paper will attempt a brief assessment of the reforms, based on relevant data on various aspects of the ESI in the UK.

3.5.1. Pro-competitive Restructuring

In the interwar period the ESI in the UK consisted of literally hundreds of small companies generating and distributing electricity. In 1926 the Electricity (Supply) Act set up the Central Electricity Board to construct and operate a national system of interconnected generating stations. That marked the beginning of efforts to centralise production and control, but was not much of a success. In the moment of nationalisation there were still 369 municipal and 200 private electricity undertakings. The latter operated mainly in distribution and were heavily regulated by the Electricity Commission, established in 1919 to promote, regulate, and supervise the supply of electricity on a national scale.

Rescue from the complications that had grown up through successive layers of regulatory legislation and inconsistency in government policy was an important motive for nationalisation (FOSTER, 1992, p. 73). In the process of transferring the industry into public ownership the Attlee government radically changed the structure of the ESI in the UK. The Electricity Act 1947 established the Central Electricity Authority (CEA), a public corporation responsible for the generation and supply of bulk electricity, and 14 Area Boards, each of them a public corporation which was to take care of the electricity distribution in its own region. Later on in England and Wales the Central Electricity Generating Board (CEGB) was created, and more autonomy was given to the Area Boards. The CEA was transformed into the Electricity Council; a federal body entrusted with a consultative and deliberative function rather than with powers of direction, control and supervision. This vertically separated structure survived until the moment of privatisation.

The only major legal change in the nationalised period, *i.e.* the Energy Act of 1983, which was supposed to reduce entry barriers and in this way introduce some competition, failed completely. The lesson to be learnt from this failure is not that the liberalisation of the ESI, even if at the starting point all the assets are publicly owned, is non-viable but that regulation for competition is needed, i.e. that new entrants have to be protected against the anti-competitive behaviour of incumbent firms. More specifically, freedom of entry is not enough if fairness in interconnections (grid access) is not assured. Prices for using transportation facilities have to be if not the same, then at least similar for incumbents, whether or not they own and/or control the grid, and entrants, and the provisions ensuring this have to be properly monitored and enforced.

The success of privatisation of gas and telecom utilities as well as the victory over striking miners in the 1980s made it possible to privatise the ESI in the UK, but restructuring had to precede selling the shares and that for two reasons: firstly, to reduce uncertainty which could strongly disrupt the running of the industry in the future, and secondly, to avoid a breach of faith with new shareholders, *i.e.* to avoid their expropriation.

Two factors strongly affected the final outcome of restructuring. Namely, the misconceived efforts to privatise nuclear power stations together with coal fired ones combined with the political pressure to complete the ESI privatisation before the next general election strongly affected the way in which the electricity generation was restructured: *competition [was] twice sacrificed: once upon the altar of nuclear power (which forced the duopoly) and secondly to the need to expedite the sale* (ROBERTSON *et al.*, 1991, p. 97). This is not; however, to deny that the UK privatisation of the ESI did represent a revolution in the economics of the industry when judged by international standards.

Instead of breaking the CEGB power stations into between five and ten groups, as suggested by some authors (SYKES AND ROBINSON, 1987,

HENNEY, 1987, ROBINSON, 1988, POLLITT, 1992), apart from Nuclear Electric, most of whose assets remained in the public sector until 1996, only two new generating companies were created: National Power and PowerGen, the market shares of which were in the moment of privatisation were equal to, respectively, 70 and 30 per cents. Of course, the pool receives electricity also from Scotland (Scottish Power, Scottish Hydro-Electric, Scottish Nuclear, and from the French EdF) and since then the market share of the two companies privatised south of the border declined substantially, but, as we shall see, the worries about some form or another of collusion remain and drive efforts to reform the way in which the pool operates.

In the process of abolishing the CEGB one more company was created, namely the National Grid Company (NGC), which was also privatised but not by public offer: initially it was jointly owned by the privatised Regional Electricity Companies (RECs) and run at arm's length, and later floated on the London Stock Exchange. In this way generation has been vertically separated from transmission. The NGC is a key player in the restructured ESI. Firstly, it is responsible for the high voltage transmission of electricity from generators to distributors and some consumers, its wires being open to common carriage by third parties. Secondly, the NGC is running the spot market for electricity (the so-called electricity pool), acting as an auctioneer and load dispatcher with authority over the operation of generating stations. The merit order dispatch is operated on the basis of bids made by generators but buyers and sellers are linked not only by the short-term spot market but also by various longer-term financial contracts (the so-called contracts for differences).

The next link of the ESI chain is distribution, in the process of which electricity transmitted through the high voltage grid gets to final consumers. The 12 Regional Electricity Companies are direct heirs of the same number of Area Boards and, in order to avoid allegedly wasteful duplication of network, they enjoy monopoly rights in their own respective territories. They are also responsible for system reliability. Distribution with its common carriage provisions should not, however, be confused with supply, which essentially consists of the contract between final consumer and supplier in which the supplier delivers electricity which it may have bought from a generator, or another regional distribution company, or which it may have generated itself.

Until 1998/99, for buyers with maximum demand less than 1 MW, and less then 100 kW after 1 April 1994, the RECs were also necessarily their suppliers, but the two roles were nevertheless separated in these companies= accounts. The supply business of any REC must purchase electricity transportation services from its own distribution business, and that on terms similar to those available to other suppliers, whose number is constantly increasing. In other words, apart from generation, supply has gradually become fully open for competition and, following a similar liberalisation of the gas market, since 1998/99 all final customers have the right to choose their suppliers.

3.5.2. Regulation for Competition

The creation of the Office of Electricity Regulation (Offer) may be legitimately considered as part of the restructuring of the ESI and the Director General of Electricity Supply (DGES) or rather his activities as another link in the supply chain (in 1999 this role has been taken over by the Office of Gas and Electricity Markets, Ofgem and its Director General). The restructuring was however a one-off action or so at least one hoped, while regulation is a process and it is by no means certain that it is going to wither away. It may however become more focused on naturally monopolistic parts of the industry, while provisions of antitrust legislation may become increasingly used.

The regulatory agencies created in the process of privatisation in the UK are non-ministerial government departments headed by a Director General, appointed by the relevant Secretary of State - in most cases it is the President of the Board of Trade, *i.e.* the Secretary of State for Trade and Industry, who after the Department of Energy has been abolished concentrates in his hands most powers regarding public utilities - for not more than five years. He may be removed only "on the ground of incapacity or misbehaviour". The regulator appoints his staff within a budget provided by the Treasury, budget over the size of which he has no control.

The duties of the regulator are specified in the respective privatisation acts as well as, in much more detail, in the licences issued to individual utilities by the Secretary of State. All regulators have a primary duty to ensure some form of universal service and to ensure that the regulated utility has the financial resources to meet the obligations imposed on it. This means that the regulator must not with his or her own actions undermine the financial position of regulated companies, and not that they are protected from commercial risks and possibly even bankruptcy. Subject to meeting these objectives the regulator has a secondary duty to take steps to increase competition, although again the wording differs substantially, to control prices, quality and service levels, to promote efficiency and economy and other duties relating to health, safety and R&D.

The cornerstone of economic regulation UK-style is the price cap "RPI-X", imposed either on prices charged for or on revenues from given types of activities. This is enforced by the regulator and modified in agreement with the industry, or in absence of this by referral to the MMC (now to the Competition Commission). The MMC acts as an "appeal court" for licence revision and re-setting of the price cap. The reference is usually a general public interest reference rather than a narrow determination of the issue in dispute, and the managers of regulated utilities are usually keen on avoiding it. One reason for that is that after the MMC report has been published, the regulator can change the licences so as to remedy the detriment to the public interest, but in principle he is not required to accept the recommendations specified by the MMC. That is why implicit regulation, *i.e.* the very threat of - or the risk of triggering - both a regulatory intervention and a referral to the MMC have themselves powerful disciplining effects and play such an important role in the British regulatory system.

Judicial review, or rather its possibility, is yet another important element of the regulatory game in the UK. However, the principles of judicial review require that a regulator's decision not be tainted with "illegality, irrationality of procedural impropriety", which is rather difficult to establish if, as is the case in the UK, the decision makers are not obliged - and usually avoid - to present any reasons for the decisions reached.

From the structural point of view the ESI regulation differs from the regulation of other utilities in that the DGES has more extensive and more explicit powers to obtain information from licensees than there are for some other directors. He also stands out in that he has a primary duty to "promote effective competition". According to the Electricity Act 1989, the Secretary of State, as well as the regulator, have duties to promote research into, and the development and use of, new techniques, to promote energy conservation and to protect the environment. Finally, the Secretary of State not only remains in charge of parts of the ESI; *i.e.* of nuclear power which remains in the public sector (Magnox Electric) but also can veto any licence modification agreed between the regulator and the company. How to make the relationship between politicians and regulators more transparent has been particularly hotly debated since the Labour Party came to power in May 1997.

How are consecutive stages of the electricity supplying chain regulated? The principle applied in the ESI in England and Wales is that one ought to regulate the naturally monopolistic parts of the industry (transmission and distribution), and, initially, supply to smaller users, and to rely on competition in generation and in supply to larger users. In consequence, the least regulated sector of the ESI is generation where, subject to planning permissions, in itself not at all a trivial issue, anybody can enter and produce electricity, using inputs of his choice, with exception of nuclear fuels. Such a new establishment has the right to be connected to the national grid and the only restriction is that any new sizeable generator must join the electricity pool.

The supply market was initially divided into two parts: franchise and contract ones. The borderline was lowered in 1994 from 1 MW to 100 kW and removed in 1998/99: following the experience with total liberalisation of the gas market, it was decided that the final stage of electricity supply liberalisation should take place at different dates in different regions. Contracted supply is, just like generation, potentially very competitive, even though possible gains from competition in terms of lower prices are definitely larger in generation, as supply accounts for only 4 per cent of overall costs in the industry.

This leaves us with two sectors of the ESI where we deal with regulation in the strict sense of the word: transmission and distribution. In both cases specific to them regulations take the form of price caps in the form of the famous RPI-X formula. All forms of regulation and in particular respective X terms are subject to regular reviews.

Apart from the famous formula RPI-X, there is one more aspect of regulation UK-style that distinguishes it from its predecessors elsewhere, namely regulation for competition, which, as distinguished from general competition and consumer protection laws, is a new invention and nowhere can it be seen better than in comparison with regulation in the US: *there is nothing in US utility regulation approaching a statutory duty to promote competition* (BEESLEY and LITTLECHILD, 1991, p. 45).

Promoting competition involves facilitating the entry of new competitors, including the entry of existing competitors into new parts of the market, which obviously requires that the regulator is aware - and able to assess - the relationship between his actions and the probability of entry. Regulation for competition was needed for a few reasons. At the most general level one could point out the very complex relationship between restructuring, competition, deregulation, regulation, and liberalisation. This relationship is both path- and environment dependent, and politics, especially politics understood as process of interaction among various pressure groups and vested interests, which differ, from each other in their relative strength, has considerable influence on specific outcomes. More specifically, the duty to promote competition reflects in part practical difficulties with moving in one go from a nationalised monopoly to a competitive industry. *However, this [regulation for competition] is a marked departure from the ideal of regulation based on a level playing field* (VELJANOVSKI, 1993, p. 55). The promotion of competition as a distinct objective of the economic regulator also creates a potential conflict of interest with consumer protection: competition can lead to reductions in benefits to some consumers, especially if they were previously receiving preferential treatment; it is also hardly compatible with many social objectives. The very concept of cream skimming shows the difficulties of squaring rivalry and universal service (VELJANOVSKI, 1993).

3.5.3. Privatisation

One more question which remains to be answered in the context of presenting British experiences with promoting (effective) competition in the ESI may be formulated as follows: was pro-competitive restructuring of the ESI in England and Wales and regulation for competition conditional on privatisation, and if yes, to what extent? Or perhaps more generally, is competition dependent on whether or not a given industry is in the private sector, and if yes, to what extent? In general it seems that privatisation is an almost necessary condition for competition understood as rivalry and for firms being really subject to what is sometimes called competitive pressure. Rivalry between state-owned enterprises (SOEs) within one industry is usually only rivalry for resources from the Treasury but not in and for the market (*e.g.* contracts of central and/or local governments).

There is no doubt, that in Britain in the 1989/90 reforms the government's commitment to privatise the ESI was a necessary pre-condition of pro-competitive restructuring of this industry and of introducing regulation for competition. But privatisation cannot solve all problems. It is true that despite the multiplicity of objectives that individual governments want to achieve by selling off the family silver, the main rationale for such actions should be that it can help to improve economic efficiency. It is also true, that the most important insight justifying this hope is that, among other things, the economic markets lead to better results than political ones. This can be seen in a particularly clear way in what is called the "market for property rights" and the "market for managers". The possible gains from depoliticisation of competition and regulation, whatever the rationale for the latter, only strengthen this insight. However, one must immediately notice the most important paradox present in the case of the privatisation of a firm in a market economy: the negative effects resulting from "government failure" can only be remedied by political means. Firstly, the very decision to transfer ownership rights has to be taken by politicians. Secondly, the decision to go ahead with the transfer is usually accompanied by deliberately introduced changes in the degree of competition and/or in the regulatory framework faced by the newly privatised companies. In most cases such changes are in fact unavoidable and this is what makes any comparison between the periods before and after privatisation almost impossible. Furthermore, these changes are discretionary in the sense that even if we assume that they aimed at improving economic efficiency, which is by no means certain (not least because of the many goals that each privatisation was supposed to achieve), such an aim did not predetermine unequivocally the option to be chosen and, as we have seen, changes in competition and in regulations are of crucial importance. But the decisions in this sphere were exposed to the same dangers (stemming from the political character of the process of taking them) which these decisions were supposed to eliminate. This is true also for the choice of the techniques of privatisation, as well as for other strategic choices. In other words, it is only politicians who can (re) establish the priority of economics over politics, but why should they be interested in doing this at all, not to mention doing it properly, if they are, as is plausibly assumed, utility maximizers? Gains from eliminating political interference and from exposing agents to economic pressures (conditional upon transfer of property rights) may be potentially large, but may remain unachievable within the political processes. And this is the last but by no means the least important lesson from the British experiences. In other words, privatisation, or rather the decision to privatise the ESI was in fact a necessary precondition for promoting competition in this industry but in the political process of putting it into practice particular attention has to be paid to the risk of expediently compromising initial commitments.

3.5.4. The Fossil Fuel Levy

Without going into the intentions, revealed or otherwise, of the then Government, at the very simplest, the Fossil Fuel Levy is collected from licensed suppliers of electricity - either Public Electricity Suppliers (PESs) or Second Tier Suppliers - and the revenue transferred to the Public Electricity Suppliers. However, it must be noted that at the beginning was the Non-fossil Fuel Obligation, on the basis of which PESs were obliged to buy a pre-specified amount of electricity produced by non-fossil fuel generators (nuclear and renewable) in order partly to satisfy the demand for electricity from their franchise customers (initially those with peak demand of less than 1 MW, from 1.04.1994 until 1998/99 - less than 100 kW). Since at that time such electricity was much more expensive than that produced from other sources (which either no longer is the case or will soon cease to be the case, if cost and price trends can be relied upon), it was decided that the PESs should be compensated for the extra costs incurred in the process of fulfilling this obligation. Therefore a levy was imposed on electricity supplied by licensed suppliers (initially 10 per of the value of supplied electricity, currently - 0.7 per cent). The levy was collected on behalf of the Director General of Electricity Supply, who was responsible for determining its level for the given period with three months notice. At the same time, in order to facilitate and simplify dealing, it was decided that the PESs would join their forces in the context of the said obligation, so that its fulfilment would be handled for them by the Non-fossil Purchasing Agency.

The way in which the levy is administered is conditional on the organisation of the electricity market in England and Wales: all electricity generated by source of more than 1 MW capacity, has to be bid into the pool (spot market), which serves as a basis for dispatch operations run by the NGC. The bidding regarding half-an-hour long periods of each day results in the so-called pool price, which in turn becomes a reference point for settlements of contracts between electricity suppliers and final customers. These contracts are known as contracts for differences (CfDs): if the pool price happens to be different from the price specified in the contract - which is most of the time - the difference is paid to the final customer (if the contract price is the higher one) or by the final customer (if the contract price is lower); that is so because both parties initially receive and pay the pool price for electricity sold to the pool and bought from it, respectively.

Contracts signed with the producers of electricity from renewable sources, selected on the basis of regular tenders - and, initially, with Nuclear Electric - are contracts for differences. Electricity from these producers Agoes@ to the pool, where it is sold at the pool price, and its producers are later paid the difference between the pool price and the price specified in their contracts. This difference was sometimes called a premium. These contracts were signed between the producers and the Non-Fossil Fuel Purchasing Agency, which acted on behalf of the PESs, who were under obligation to purchase electricity produced in this way. So, effectively, the money collected from all licensed suppliers ended up with renewable and nuclear producers - via, *de facto* PESs, but the chain, as we have seen, is longer and included the collectors and the DGES and the purchasing agency. Since the contracts were between producers and the purchasing agency, acting on behalf of the PESs, one could say that the PESs themselves did not see any of that money, and - without going into unnecessary details - that the levy was used to pay the premium, mentioned above.

Regardless of whether the name of the levy was only a fig leaf for a rescue operation of nuclear industry in the UK, there is no doubt that gross of the money - about 1 billion pounds a year - ended up with the nuclear generators. Their organisational structure changed over time. Initially, that is in 1990 two publicly owned companies were created: Nuclear Electric in England and Wales and Scottish Nuclear north of the border. In 1996 the structure based on location was replaced by one based on technology: power stations with AGR and PWR reactors have become parts of British Energy (privatised in June 1996), and those with older Magnox reactors - of Magnox Electric, later merged with BNFL, plans for privatisation of which were announced in July 1999. There is no doubt that the money from the levy helped the nuclear industry in England and Wales not only to survive but also to achieve commercial viability - increases in efficiency of nuclear power stations in Great Britain have almost miraculous dimensions - although there is no doubt either that the way in which nuclear power stations were sold was instrumental in avoiding the so-called stranded costs (JASINSKI, 1998). At the same time the money from the levy was used to support and indeed hugely develop generation of electricity from renewable sources (ROSS, 1998). This was achieved in a way fully compatible with newly introduced competition in the ESI in England and Wales, and the competitive process of allocating support resulted in reducing average price of so produced electricity from 4.35 p/kWh in 1994 to 2.71 p/kWh in October 1998!

3.5.5. Assessment: Market and Prices

The measures taken and to be taken to introduce competition into the ESI in England and Wales constituted a programme the implementation of which was to take eight years, and actually took almost a year longer because of delays in removing the last remnants of the franchise market. For some RECs that was done only a few months ago, and therefore one will have to wait for quite a while before it is possible exhaustively to evaluate the performance of the English and Welsh market for electricity.

Having these *caveats* in mind, one could nonetheless ask, whether the market for electricity in England and Wales is a functioning one, and that functioning well? This question, however, is far from being a straightforward one and leads to many others. For example, what is a well functioning market in an industry with naturally monopolistic elements? And how can one spot it and/or distinguish it from not-so-well-functioning markets? Is it enough to observe firms competing with each other, however difficult it may be unequivocally to interpret their behaviour in this way, or do we have to approve the effect that their rivalry has for consumers and for the economy as a whole? In other words, do we have to like the results and how patient do we have to be in waiting for the expected ones? Is not it enough to be convinced that the principles of competitive behaviour are not violated?

Assuming that competition should eventually become a self-sustaining market mechanism, we shall look from this point of view at the activities of the Director General first of Offer and now of Ofgem. An important element in assessing the competitive electricity markets is also behaviour of prices and costs over time, both in themselves and in relation to other sources of energy, changes in fuels used for electricity generation and atmospheric emissions.

In this context it is worth stressing that the regulator has no power explicitly to control generation prices and therefore prevent anti-competitive pool prices being bid by the large generators, if this was the case. Nevertheless, movements in prices have been attracting quite a lot of publicity and in consequence the regulator usually felt that he had to react. Many instances of controversy resulted in a "voluntary undertaking" by the generators to bid in such a way that prices did not exceed certain level and in them - again "voluntary" agreeing to sell some of their generating capacity, and that twice: in 1994 and in 1998/99. In the former case, *Financial Times*'s Lex Column described that development as an expression of Prof. Littlechild's curious fondness for blunt engineering solution: *Faced with the complex but failing market in electricity generation, the regulator has finally settled for giving it a couple of healthy bangs with the hammer* (12/13 February 1994). A representative of the Association of Independent Electricity Producers said immediately after the announcement: *We do not understand why he [Prof. Littlechild - PJ] thinks that artificially depressing prices will encourage competition* (in: *Financial Times*, 12/13 February 1994). These conflicts translated over time into an increasing level of dissatisfaction with the way that the spot market operated especially from the point of view of large energy users. This led to attempts to reform the electricity trading mechanisms, but if no consensus appears; primary legislation will be necessary, but difficult to assure. Other controversies regarded reopening the distribution review in 1995 and the so-called "dual fuel" deals - joint marketing of electricity and gas, which became a problem because of different liberalisation timetables for the two energy sources.

What does it all tell us about how well competition works in the ESI in England and Wales? This market for electricity has been all the time is a market under close supervision. Market mechanisms, to the extent that they have already been introduced are not yet fully trusted. In consequence the Director General of Offer is under pressure to react or at least to comment on almost any movement (read: increase) in profits of the companies involved and/or in prices charged to customers - and almost everybody is a customer. It still also a very politicised market, and the change of government in May 1997 did not change much. The Labour Party came to power promising a far reaching reform of the regulatory structures but the only change introduced so far consisted in merging Offer and Ofgas, which looks like a good idea, but from the point of view of the mechanics of regulation is rather cosmetic. Another change introduced since 1997, *i.e.* the gas fired power stations moratorium, was purely political and decisively opposed by both regulators. In the whole period not without importance were also multiple mergers and take-overs, which led to continuous changes in the industry structure and appearance of the so-called multi-utilities, of which the best example is Scottish Power, active in electricity, gas, telecoms and water industries.

The effects and consequence of liberalisation of the ESI in the United Kingdom - admittedly, in its most radical form limited to England and Wales - can be looked at from a few different points of view (the remainder of this section is based on various publication by the Department of Trade and Industry, and most Figures and Tables come from *UK Energy in Brief*, December 1998). Firstly, the newly acquired commercial freedom of electricity producers could be expected to have led them to reassess their use of various fuels and therefore result in changes in fuel use changes in electricity generation. This took two forms in the UK: a rapid increase in use of natural gas and an equally rapid decline in use of hard coal. The rate of the latter is much higher than of the former as one had to add increase production of nuclear power station and more effective support for renewables. This is illustrated in Figures 11 and 12 and Tables A14 and A15, where the data for the period 1970 - 1990 were added in order to be able better to assess changes in trends.

In the early 1970s the advent of natural gas saw gas consumption grow rapidly. In the last 20 years industrial consumption has been relatively static, growing by 5%, while domestic consumption has grown by 80% and services consumption has more than doubled. However, in the last few years the growth in gas use has been dominated by its use for electricity generation, which increase almost thirtyfold. Defying world trends in that respect, newly constructed gas fired power stations are used in England and Wales mainly to produce base load electricity. Electricity generation now accounts for 26% of natural gas consumption, and although there are still new CCGT power stations under construction, there is currently in force a moratorium on new power stations. As we have already mentioned, this moratorium proved highly controversial and is likely to be challenged in the EU institutions.

Coal production was 3% lower in 1997 than in 1996 due to a 6% fall in deep mined production, while opencast production rose by 22%. Coal production in 1997 was only a third of the level in 1970. The deep in 1983-4, in Fig. A2 was due to a prolonged miners-strike, but much more important it is to notice that the downward trend in coal production started much earlier is continuing and expected to continue in the foreseeable future. This is mostly due to changes in coal consumption.

It is true that in the years 1990 - 1997 power stations reduced their coal consumption from 84.0 to 47.1 million tonnes, but at the same time they accounted for 75% of coal consumption in 1997 compared with 78% in 1990, 73% in 1980 and 49% in 1970. Coal consumption has declined more sharply during the 1990s; over the last six years at an annual rate of 82% compared with 12% per year over the previous 20 years. It can therefore be said that putting blame for the long-term coal industry demise solely on the electricity liberalisation is simply wrong.

To changes in the use of coal and gas in electricity generation, one has to add changes in the output of nuclear power stations and in the use of renewable energy sources. The relevant data are included in Figures A4 and A5 and Tables A17 and A18.

The British nuclear power generation industry achieved record output of 89 TWh in 1997 as well as contributing, at 28%, its highest ever proportion of total electricity generation. Electricity output was over 50% higher in 1997 than in 1990, although in that period only one new nuclear power station - Sizewell B - was commissioned. It follows that most of the increase was achieved thanks to higher efficiency, which in turn can be traced back to the desire of the then management to have their company privatised. In order to achieve this objective it was necessary to prove that electricity generation in nuclear power stations was commercially viable without any - hidden or otherwise - subsidies.

Biofuels account for 82% of renewable energy sources with most of the remainder coming from large-scale hydro electricity production. Wind power contributes 22%. Of the 2.3 million tonnes of oil equivalent of primary energy use accounted for by renewable, 1.4 million tonnes was used to generate electricity and 0.9 million tonnes to generate heat. Renewable energy use has doubled since 1990.

The data contained in Figures A1 - A5 and Tables A14 - A18 (appendix) can be summarised in tracing changes in fuel use for electricity generation - see Fig. A6 and Tab. A19, where the data for the years 1970 - 1990 were again added for comparison. On the basis of these data one can conclude that the fuel used to generate electricity has changed markedly in recent years. While coal still has the largest share of the market for electricity generation, the proportion has fallen from two-thirds to under two-fifths in seven years. In 1997 gas accounted for 27% of the market for electricity generation. This dash for gas, as it is usually referred to, led recently to worries about too high a level of dependence on this fuel as well as that it would deliver a final blow to the UK coal industry. That is why the current Labour government decided to introduce a highly controversial two-year moratorium for gas fired power stations.

Changes in generation of electricity have to be confronted with changes in its consumption as well as with changes in prices, both in absolute and relative terms. The relevant data are contained in Figures A7 and A8 and Tables A20 and A21. As one can see, over the last years electricity consumption in the domestic and services sectors has grown by 5% and 15% respectively. Industrial consumption varies with business activity, but has risen in each of the last three years to be at its highest ever level in 1997.

The issue of prices attracts a lot of attention in all discussions on potential and actual liberalisation of the ESI in any country, and the UK is no exception. At the same time one consequence of liberalisation is that it is increasingly difficult to talk about the price of electricity, even for a given group of customers, as for example, the obligation to publish tariffs gradually disappears in line with effective competition taking roots. At the same time one has to remember that the highly publicised pool prices are only very indirectly related to what anybody actually pays. That is why one way of looking at what happened to electricity prices in the UK is to compare them with prices of other fuels for the industrial sector, the data for which are provided in Fig. A8 and Tab. A21.

Industrial electricity prices fell sharply, by 8 per cent in real terms, in 1997 compared to 1996, and are now 212 per cent lower than they were in 1990. One factor behind the fall in the latest year was the full pass-on of reductions in the Fossil Fuel Levy. Average industrial electricity prices are now lower in real terms than for any year since records began in 1970. The same is true for coal, where prices have fallen by 36 per cent since 1990. Gas prices rose by an average of 2 per cent in 1997 (although some larger consumers saw much larger increases), but remain at historically low levels, 46 per cent lower than in 1990. Heavy fuel oil prices fell by 7 per cent between 1996 and 1997 as crude oil prices fell.

Overall total domestic energy prices including VAT fell in real terms for the third year running in 1997. Average real prices in 1997 were 52 per cent lower than in 1996. VAT was cut from 8 to 5 per cent from 1 September, which accounts for about 1 per cent of the overall fall. Within the overall movement, average electricity prices fell by 7 per cent, heating oils by 5 per cent and coal by 12 per cent. The roll out of competition continued in gas throughout 1997 contributing to the 32 per cent fall on 1996, although consumers who have moved to new suppliers can have seen reductions of up to 20 per cent. Between 1990 and 1997 real prices have fallen by 92 per cent for electricity, 12 per cent for gas and 24 per cent for heating oils.

Last but certainly not least there are worries about the effect of liberalisation on the natural environment. As one could expect, since the UK liberalisation resulted in an increased use of natural gas and renewable sources in electricity generation and - possibly less directly - in an increase of efficiency of nuclear power stations, the contribution of this sector to CO_2 emissions diminished. This is confirmed by the data in Fig. A10 and Tab. A23.

4. The Policy Framework in Europe and Germany

4.1. EU Energy Policy and Germany´s Ecological Tax Reform

EU Energy Policy

The main aim of EU energy policy is to achieve security of supply and protecting the environment while maintaining international competitiveness. Each Member States' energy policy must be evaluated in terms of these objectives and, in addition to this, with compatibility with the single market as well as the EU's international climate policy.

The objectives of EU energy policy were clearly stated in the White Paper on Energy Policy for the European Union (COM, 95, 682 final). According to this paper EU energy policy is based on deregulation and market integration; government intervention should be efficient and effective, it should be in the public interest and contribute to sustainable development, consumer protection and economic and social cohesion. The Energy Council has adopted the Multiannual Framework Programme for actions in the energy sector in the period 1998-2002 (COM,98, 607, final). A core idea of this programme is to co-ordinate the manifold energy policy activities and initiatives of the Community, to enhance transparency and increase policy efficiency. Funding allocated to this policy field is very modest given the ECU 170 million agreed upon by the Energy Council in November 1998. Additional funds could be allocated to projects in cohesion fund countries and to all Member States to the extent that the Commission assigns the 0.5% from the Structural Funds reserved for innovative projects (here in the field of energy policy).

The EU has emphasized some structural objectives:

- increasing the share of renewable in energy generation – if possible doubling this share to 15% by 2010;
- raising the share of natural gas which is rather environmentally friendly as an energy input;
- achieving maximum safety standards in nuclear energy;
- maintaining the percentage of solid fuel (hard coal and lignite) in total energy consumption figures which seems to reflect the desire of some countries – the UK, Germany and Spain - to maintain the medium-term competitiveness of their coal fields. One must, however, point to the inconsistencies of coal subsidization in several countries (GREENPEACE, 1997).
- There is also some EU financed research on energy conservation, the development of alternative energy sources, nuclear fusion and other projects.

With respect to the internal market the Community adopted two important Directives. As a follow-up on Directives of 1990 and 1991 on transit of electricity and gas a considerable opening of the electricity markets – including international trade in the context of third party access to the electricity network – was agreed upon in the Directive 96/92/EC (Official Journal 30.01.1997). Moreover, the Commission and the European Parliament agreed on a common position on a Directive (98/30/EC) which liberalized the gas market in 1998. The European Commission is reporting once a year on progress in the field of the two Directives.

The Community has an action plan for cutting greenhouse gases – especially CO_2. This policy field became crucial in the context of the UN Kyoto conference of 1997 and other international conferences. According to the EU's commitment it must cut CO_2 emissions by 8% of the 1990 level by 2008-2012. The Community is facing additional challenges in the context of EU enlargement, which will consist of at least two waves of new member countries. The European Energy Charter – ratified in late December in 1997 – defines a framework for east-west cooperation in the field of energy policy.

The electricity output of the EU increased between 1997 and 1999 (EUROSTAT, 1999): Comparing 1998 and 1997 there were strong increases in Portugal (+14.9%), Greece (6.5%), Belgium (+6%), Ireland (5.9%), Sweden (+5.4%), the Netherlands (+4%) and Italy (3.3%). Production increases were modest in the UK, Finland, Austria, France, Spain and Germany, negative growth rates were observed in Luxembourg (-8.3%) and Denmark (-6.5%). With respect to the structure of electricity production one may notice that the share of conventional thermal power generation has increased (+4.6%) and accounted for 51.5% of total production; electricity from nuclear power stations fell slightly and represented 34.1% of total production. Hydroelectric power generation increased by almost 5% and represented 14.4% of the total. In 1998 there were EU net imports of 13.8 TWh which accounted for 5.9% of production. France, Germany, Denmark, Austria and Sweden were net exporters; the largest importers were Italy and the UK. Total electricity supplied in the Community was 2329.5 TWh, an increase of 2.1% compared to 1997. As regards the structure of electricity sources only a few EU countries recorded shares for hydroelectric (and other, non nuclear and non thermal) power of more than 20% in 1998: Spain, Italy and Finland – each with a share of 1/5 -, Portugal and Sweden with 34.5% and 47.7%, respectively; the share in Luxembourg was 90.9%, where one may note that Western Europe's top country is Norway with 99.4%.

Germany's federal government has emphasized the goal of energy efficiency policy for many years, and there are fiscal measures as well as regulations in the transport and building sector. Municipalities and state governments finance a host of measures to improve energy efficiency, in particular by giving loans and grants to the business community. All those measures must be notified to the European

Commission. However, there is no real transparency about energy conservation expenditures as one can learn from the OECD (1998) report on Germany's energy policy – a very unsatisfactory situation (in a leading industrialized country) which urgently should change on the basis of mandatory full reporting of all municipalities and states. The OECD notes (p. 42/43)

"There are no legal or constitutional restrictions on the measures related to energy efficiency which local governments are empowered to take, but as a general rule the same project is not allowed to receive funds from different public authorities at the same time. Some Laender such as North-Rhine/Westphalia and Saxony are more proactive in the energy efficiency field than others. They have no obligation to report to the Federal government on their activities. However, the Federal government seeks to be informed. A survey on the use of Federal government and Laender funds was completed in 1997. This survey did not take into account the expenses of the municipalities and of some Laender, which did not respond to the survey. Table 10 indicates the result of the survey: total expenditure from the Laender who responded to the survey has been rising and in recent years has become comparable to that of the Federal government."

Tab. 10: Energy Conservation Expenditure in Germany, 1990-96
(Federal Government and Some Laender)

million DM

	1990	*1991*	*1992*	*1993*	*1994*	*1995**	*1996**
Federal Expenditure	1 242	1 008	1 212	1 186	1 285	1 316	1 461
Laender Expenditure	556	978	802	1 092	1 604	1 169	956
Total	1 798	1 986	2 014	2 278	2 889	2485	2 417

* Provisional

Source: Country Submission

EU Proposals for a CO2/Energy Tax

The European Commission made two draft proposals for a CO_2/Energy Tax in 1992 and 1995, however, it was only in 1997 (March 12) that the Council of the European Community presented a Council Directive "Restructuring the Community framework for the taxation of energy products"(COM, 97, 30 final). In line with an earlier German proposal – made in the context of Germany's EU presidency in 1994 – it was proposed to increase the minimum tax rates for mineral oil and to extend the scope of mineral oil taxes to all energy products like gas, solid energy products (hard coal and lignite) and electricity. The proposed Directive would mean than a new tax regime would enter into force as of 2002 provided that the European Parliament had agreed to this in its deliberations. Existing EU Directives on the taxation of mineral oil products (92/82/EEC and 92/81/EEC) thus would be replaced. The proposed Directive suggests several tax exemptions – see Articles 13-16 – and basically indicates several types of energy use which would be subject to different tax rates: motor fuels, motor fuels for certain industrial and commercial purposes (e.g. agricultural use), heating fuels and electricity. Particular important exemption clauses stipulate that (i) energy products used as raw materials are not taxed; Member States may apply exemptions or tax reductions to renewable energy forms and to heat generated during the production of electricity (art.14). (ii) Member States may refund part of tax revenues paid by a firm with non-transport-related energy costs exceeding 10% of total production costs. Member States must refund a certain amount of the tax revenue if the energy costs exceed 20%. Applying this measure the net amount of tax paid shall not be less than 1% of the firm´s value of sales (art. 15). (iii) Member States can be authorized – for specific policy reasons and for a certain time period - to apply exemptions or level of taxation below the minimum level specified in the Directive (art. 16).

A critical evaluation of the 1997 Directive proposal is given by a report of the WUPPERTAL INSTITUTE (1998, p.52): By achieving homogeneous minimum tax rates on all energy sources to all Member States, the CEC is trying to achieve a level playing field for all competitors. (...) A crucial point is the taxation of solid energy sources, in particular coal. Since Member States like Germany, France and Spain still subsidise their coal industries; the taxation of coal seems to be a paradox. But from an environmental point of view, a starting tax rate of ECU 0.2 per GJ on coal seems to be by far too low, coal being the fossil fuel with the highest CO_2 emissions per energy unit. To tax gas, a relatively "clean" fuel, at the same rate as coal or peat does not make sense from an environmental point of view, either. (...) An electricity end use (output) taxation is easy to administrate but does not consider the varying environmental impact (external effects) of different primary energy sources used as input for electricity generation which some Member States would like to take into account. As a kind of compromise the CEC authorised Member States to tax the primary energy input in electricity generation in addition to the end use tax of ECU 1 per MWh in 1998. But while not setting minimum rates for an input taxation, it seems unlikely that Member States will impose additional taxes to reflect the environmental impact of different ways of electricity generation. Electricity is an easily tradable homogeneous good with rather low transport costs, and electricity trade in the EU is growing. As long as no harmonized tax levels for input taxation are provided, a unilateral (double) taxation is one or more Member States could distort competitiveness.

One should also emphasize that Germany's Environmental Expert Council advocated differentiated taxation of primary energy inputs.

The above statement adequately summarizes the most important points – except for one caveat, namely that differential environmental impacts of various energy inputs in electricity generation could be taken into account by a rebate system in which government would apply a rebate for a (minimum) share of environmental friendly inputs. The German Government developed a concept without much differentiation on the input side of the electricity sector. This sector will face both adjustment pressures from higher ecological taxes, which should raise prices, and from EU deregulation which will exert downward pressure on prices.

EU Liberalization Directives and Energy Policy in Germany

The EU electricity liberalization Directive came into force on February 19, 1997 and requires that national law be adjusted within two years. The required minimum opening up of national electricity markets is 23% by 1999, 28% by 2000, 33% as of 2003. A similar EU liberalization Directive was adopted for the natural gas market, which came into force in June 1998 (to be adopted into national law by June 2000). The three liberalization minimum steps require opening up for 20% by mid-2000, 28% by mid-2003 and 33% by 2008. Liberalization of the natural gas market is linked with the electricity market in the sense that the structurally rising share of gas in electricity liberalization leads to secondary liberalization effects in the power market. In the single EU energy market emerging in the coming decade one may expect a strong rise of electricity trading across borders. Germany's electricity imports reached only 7% in 1997. Intra-EU trade in electricity is likely to benefit both from the politically imposed opening-up and the ongoing international mergers and acquisitions in the energy sector. For example, the French EdF recently acquired London Electricity and also is eager to buy into the German electricity sector, a development which could be accelerated by industrial electricity prices falling below the French EdF domestic price in the wake of electricity deregulation in Germany. The Finnish IVO bought Stockholm Energi; one may also note that some British companies have acquired stakes in the US, which means the reversal of a process of US power generators investing in the UK energy market in the early 1990s. However, there are also national merger projects, e.g. in Spain, the UK and Germany, because there are considerable economies of scale in traditional energy generation – not in solar energy, of course.

Since energy is a very crucial input in all sectors and for all households, security of supply and full operability of the distribution network are crucial for OECD countries. Given the critical problems of nuclear energy generation – especially in the field of nuclear waste storage – many EU countries have considered phasing out nuclear energy (e.g. Germany and Sweden) or renouncing building nuclear power stations (e.g. Austria). The question arises as to what extent there are economically viable and ecologically acceptable alternative ways of electricity generation. Since the EU has adopted a liberalization scheme for both electricity and gas there is no doubt that competitive forces will shape the substitution process. Experiences from early liberalization countries such as the UK or Sweden, Norway and Finland suggest that gas will play an increasing role in electricity generation and that combined heat and power offers interesting opportunities in terms of efficiency gains.

Sustainable development requires that OECD countries should reduce the specific use of energy in the long run. EU liberalization of energy markets basically will bring about efficiency gains and falling energy prices, which could raise emission levels critically in Europe. Therefore it will be important to adopt a consistent energy tax policy encouraging continuous energy saving, and such a tax policy will mean taxing CO_2 and SO_2 emissions to some extent. In the long run one could combine certain national tax schemes with international emission trading schemes.

Since the EU has envisaged two stages of eastern EU enlargement and since several east European countries – including Russia - have a comparative advantage in electricity generation and energy exports, respectively, a future energy strategy of the EU should adopt the principle of open energy markets: in the long run eastern European countries could export more energy and electricity, respectively, so that modernization investment in postsocialist countries – relying often on EU imports of investment goods – could be financed from the revenues from energy exports. If EU countries should adopt taxation of energy inputs there will be complex problems of imposing adequate tariffs on electricity imports.

Modernization investment and expansion of electricity in the Baltic countries and in the Visegrad countries and Russia have allowed to create a pan-European electricity network in 1999 so that the former East-West division no longer is exists. This means favourable prospects for rising international trade in electricity.

Germany's energy policy is characterized by several elements:

a) The natural gas and the electricity sectors in Germany are characterized by a fairly large number of enterprises and by the activities of municipal companies. Competition has been impaired by concession contracts with municipalities and by demarcation agreements. While demarcation agreements are used by potentially competing companies to agree in a contract to refrain from market entry in a certain area (this private contract between two companies does not preclude other newcomers from market entry) concession contracts are the basis for municipalities to collect concession fees from electricity and gas distribution companies.

b) Typically the municipality concludes an exclusive contract with a utility company which will have the right to use the territory of the municipality to install lines for the supply of end users – sometimes municipal companies will also be involved in local distribution to end users. The system of demarcation and concession contracts applied to both gas and electricity. Demarcation agreements and concession contracts were allowed under the Act against Restraint of Competition of 1957, however, within a sweeping deregulation these exemptions from the general principles of competition were removed in the 1998 amendment to this act.

Municipalities used concession fees to cross-subsidize deficits in the other fields, mainly in public transportation. In 1999 there still were some 900 municipal companies involved in natural gas and electricity distribution and in electricity production. New laws which came into force in April 1998 eliminated exclusive concession contracts and demarcation agreements, indeed electricity and gas markets were liberalized in one go to all consumers, including households. A strong deregulation drive characterizes the electricity market where prices came down under competitive pressure in 1999. Price reductions had started earlier because of a decision from the Constitutional Court:

c) Germany has been among the few EU countries, which have adopted energy taxes in the context of an ecological tax reform in the 1990s. As Germany accounts for ¼ of EU gross domestic product and as the country is the EU's leading exporter it will be quite interesting to analyze the effects of alternative ecological tax reform concepts. Germany thus is treated as a potential role model for a broader EU ecological tax reform. At the same time the analysis can focus on the question to which extent a large economy can pursue an isolated national ecological tax policy without major negative international effects. The following sections take a closer look at selected theoretical, empirical and policy issues.

Traditionally Germany's energy policy has relied on voluntary agreements and government regulation in the field of energy generation and energy conservation. Besides federal regulation, state (Laender) governments and municipalities were involved in CO_2 emission reductions, mostly via energy efficiency programmes and incentives for renewable. The Association of Municipal Companies declared in March 1996 that they would reduce their CO_2 emissions by ¼ between 1990 and 2005. This declaration is part of the overall voluntary agreements involving the industry in Germany.

Voluntary agreements with respect to CO_2 emission reductions were made by 14 industrial sectors, four associations from the BDI (Federation of German Industries) and the Federation of German Industries in March 1995; they agreed on voluntary agreements to cut CO_2 emissions and to increase energy savings. The sectors involved represent some 2/3 of industrial energy consumption. Individual commitments were signed by industries – partly including firms located in eastern Germany – and the commitment was to achieve up to 1/5 reduction between 1987 and 2005 of specific CO_2 emissions or specific energy consumption (sometimes absolute reductions targets were also specified). German industry sharpened its commitment in an updated version of the voluntary agreement in March 1996 when they declared that they would reduce specific CO_2 emissions by 20% between 1990 and 2005. The so-called voluntary agreement programme is supported by low interest loans from the KfW (Kreditanstalt für Wiederaufbau that is a state-owned bank) and by the Deutsche Ausgleichsbank, which will finance mainly energy efficiency investment in small and medium-sized companies. Note, however, that specific energy consumption targets have an obvious disadvantage, namely that growth in national and international demand allows the raising of absolute output and emission levels. Having achieved the voluntary agreements government decided to postpone regulatory steps in the field of energy efficiency of heating systems, moreover the federal government announced that industrial sectors participating in the voluntary agreements would be exempted from energy/ CO_2 tax or that this tax would be offset as long as the respective industries met the commitments made.

Three Stage Ecological Tax Reform

The government decided to adopt a three-stage energy tax reform. The first stage came into force on April 1, 1999 when an energy tax levy was imposed on electricity, fuels, heating oil and natural gas. A new element in the German tax system is the introduction of an electricity tax where no differentiation with respect to the various primary energy inputs is applied. There are certain exemptions and reductions for industry. The estimated annual revenue of DM 11.3 bn annually slightly falls short of the envisaged reduction of social security contributions, namely 12 bn in 1999.

One has to distinguish between the standard tax rate - applying to private households, small electricity users and in the field of transportation – and the reduced tax rates for the producing sector (industry plus service sector plus agriculture and forestry).

Tab. 11: Tax Rates and Expected Tax Revenue in Stage I of Ecological Tax Reform, Germany

Type of Energy	Standard Tax Rate	Reduced Tax Rate (for Producing Sector)	Tax Revenue for Period April-Dec. 99
Fuels	6 Pfennig/l (increase)	6 Pfennig/litre	DM 2 800 million
Heating oil	4 Pfennig/l (increase)	0.8 Pfennig/litre	DM 1 000 million
Natural gas	0.32 Pf./l (increase)	0.064 Pfennig/litre	DM 1 400 million
Electricity (trams and busses) (existing heating)	2 Pfennig/kWh 1 Pfennig/kWh	0.4 Pfennig/kWh 1 Pfennig/Kwh	DM 3 300 million
		Basic tax payment: 50000 kWhx2Pf/kWh =DM 1000	
			DM 8 5000 million (=DM 11.3 bill. p.a.)

Source: German Ministry of Finance

The expected tax revenue of DM 300 million p.a. from renewable energy sources will be fully used for financing further investment in renewable energy generation. This element of the tax reform together with the energy-saving incentives from the rise of energy prices is the environmental improvement effect of the ecological tax reform – i.e. a (first) dividend for society. Note that combined heat and power generation with an annual efficiency ratio of 70% or more will not have to pay taxes otherwise imposed by the mineral oil tax. The second main aspect of the ecological tax reform (the second dividend) is that revenue will be used to reduce social security contributions by 0.8 percentage points: down from 20.3 to 19.5 points. This reduction in social security contributions will be equally shared between employers and workers/employees which – in a legal sense – both contribute to social security financing (in early 1998 the overall contribution rate reached 42.3%). With the second and third steps of the ecological tax reform the social security contributions are expected to be reduced to below 40%.

The producing sector will benefit from a rebate model for firms and sectors, respectively, which pays relatively high additional taxes under the rules of the ecological tax reform: All firms face a basic tax burden in the context of electricity use which means that 2 Pfennig/kWh are applied to the first 50 000kWh consumed, i.e. there is a basic payment of DM 1000 p.a. For every kWh above this limit, the reduced tax rate is applied, where firms can get a rebate if the ecological tax payment exceeds a critical ratio relative to the firm's savings in social security payments. The critical ratio is 1.2, that is firms exceeding this ratio can obtain from the Customs Authorities the (corrected) excess payment. To give an example: The additional energy tax payment for a firm in the context of the ecological tax reform is DM 1.5m, it benefits from the reduction of the employer contribution rate (minus 0.4 percentage points) – the reduction of social security contribution payments is DM 1m. The ratio 1.5 m/1m = 1.5 so that the threshold ratio of 1.2 is exceeded. The excess payment which can be recouped from the customs authorities is DM 1.5m minus 1.2m = DM 0.3m. This rebate system was introduced as the European Commission indicated that the initially envisaged plan to exempt all energy intensive sectors – defined by a share of at least 6.4 percent of energy costs in total costs- would violate Community law because it represented a subsidy/financial aid to certain industries for which the Commission might be hesitant to give its ok.

The German federal government has decided on June 23, 1999 that the next stage of the ecological tax reform for the period 2000-2003 will witness an increase of fuel tax rates by 6 Pfennig/l and of electricity tax rates of 0.5 Pfennig/kWh – with the producing sector paying again only 1/5 of the standard rate. The expected additional tax revenue of DM 21bn in the period 2000-2003 will be used to reduce the social security contribution rates by 1-percentage points. The first and second stages of the ecological tax reform thus reduce the contribution rates by 2 percentage points.

Manufacturing industry accounted for 26% of final energy use in Germany in 1997, down from 31% in 1985. Households and transportation recorded 30% and 27.8%, respectively, in 1997, while the service sector accounted for 16.2%. The largest share increase was for transportation the figure for which increased by roughly 10 percentage points in the period 1985-1997. The ecological tax reform will impose a relatively large burden on transportation and private households. The overall tax incidence will largely fall on private households, however, the expected price level increase of 0.2 percent (RWI, 1999) in the first stage of tax reform is unlikely to trigger aggressive wage bargaining. There is no doubt that private households will bear the burden of the new ecological tax reform; but they also will benefit from positive welfare effects of an adequately designed tax reform. It is unclear to what extent the ecological tax reform will stimulate energy-saving investment and innovation in the long run, however, the experience of the OPEC shocks in the 1970s leads one to expect favourable dynamics.

In Germany, industry was resistant to an energy tax and declared that self-commitments of industry would be an adequate policy strategy. In the early 1990s industry wanted to convince government to give up the idea of any broad energy tax. Indeed, the German government decided to postpone regulatory and fiscal measures and decided to support the "self-commitment strategy" at the EU level. However, one may anticipate that with an efficient split of ecological tax revenues, a considerable part of Germany's industry would be in favour of a phased ecological tax reform. To the extent that an innovation-oriented ecological tax reform would strengthen the system of markets while stimulating Schumpeterian dynamics the envisaged modified tax reform would reinforce the market economy system. It is obvious for many reasons that nominal corporate and labour tax rates should be reduced in Germany (LOEFFELHOLZ, 1999).

The German government's energy policy mainly relied for many years on regulatory policy, which defined certain technological minimum standards. While the large scale combustion plant ordinance sets maximum standards for large facilities, the small scale combustion plant ordinance defines maximum heat losses and pollution levels for combustion facilities with heat output of less than 1 MW – in the case of solid fuels used -, less than 5 MW – case of liquid fuels – and less than 10 MW (gaseous fuels). This ordinance was tightened in November 1996 when further required reductions with respect to heat losses in new oil and gas-fired heating systems were set. As regards existing plants heat losses have to be reduced to certain threshold levels after a grace period.

4.2. Modelling an Innovation-Augmented Ecological Tax Reform

We base our further argumentation for a European ecological tax reform on a CO_2 tax, which lies on every energy carrier in relation to its carbon contents. But only those energy inputs are taxed, which are used for power generation. We take for our further discussions the carbon tax without any exceptions, because this is from a theoretical point of view the ideal taxation. Many arguments for exceptions from this general principle for specific sectors or energy carriers, such as we just described for the German example and that can be found in other national solutions, will be more or less obsolete in the case of a EU wide ecological taxation.

We will discuss the effects of such a European carbon tax according to different ways of using the tax revenue. MEYER/ BOCKERMANN/ EWERHART/ LUTZ (1999) have shown in their simulation studies with the model PANTA RHEI, that a carbon tax can reach the environmental targets and have positive effects on employment, if the tax revenue is used for a reduction of social security payments. But a reduction of GDP-growth is unavoidable. This result is also met in simulations with other models (SCHMIDT/ KOSCHEL, 1999; WELSCH, 1999). Our question is, which effects can be achieved for the German economy if we generally reduce the social security payments but use a smaller part of the tax revenue for support of R&D expenditures of firms.

The instrument of our analysis is the model PANTA RHEI, which is an ecologically extended version of the 58 sector econometric simulation and forecasting model INFORGE (Interindustry Forecasting Germany) (MEYER/ EWERHART, 1997; MEYER/ EWERHART, 1998). Its performance is founded on the INFORUM philosophy (ALMON, 1991), which means to build econometric input-output models *bottom up* and *fully integrated*. The construction principle *bottom up* states that each sector of the economy has to be modelled in great detail and that the macroeconomic aggregates have to be calculated by explicit aggregation within the model. The construction principle *fully integrated* means a model structure that takes into account a variable input-output structure, the complexity and simultaneity of income creation and distribution in the different sectors, its redistribution among the sectors, and its use for the different goods and services the sectors produce in the context of globalizing markets. In this way one succeeds in describing properly the role of each sector in interindustry relations, its role in the macroeconomic process as well as its integration into international trade.

These conceptual advantages end up in a consistent and powerful processing of sectoral and macroeconomic information. The about 30,000 equations of INFORGE describe the interindustry flows between the 58 sectors, their deliveries to personal consumption, government, equipment investment, construction, inventory investment, exports as well as prices, wages, output, imports, employment, labour compensation, profits, taxes, etc. for each sector as well as for the macro economy. In addition the model describes income redistribution in full detail. The model frequency is annual; the model updating frequency is semi-annual.

PANTA RHEI additionally is equipped with a deeply disaggregated energy- and air-pollution-model, which distinguishes 29 energy carriers and their inputs in 58 production sectors and households as well as 8 air pollutants (CO_2, SO_2, NO_x, NMVOC, CH_4, CO, N_2O) and their relations to the 29 energy carriers. Energy demand is fully integrated into the intermediate demand of the firms and the consumption demand of the households.

INFORGE/ PANTA RHEI are part of the INFORUM International System (NYHUS, 1991) that links 13 national I-O models on the sectoral level via export and import flows as well as the corresponding foreign trade prices. The information gain of this system in comparison to isolated models allows for a reliable analysis of the important contribution of exports for the performance of the German economy. The International System forecasts the economic development of Belgium, Germany, France, Great Britain, Italy, the Netherlands, Austria, Spain, USA, Canada, Mexico, Japan, and South Korea in full sectoral disaggregation. This world trade model is being developed steadily, in the near future models for China, Taiwan, and Poland will be integrated into the system (MA, 1997; NYHUS/WANG, 1997). Besides the goods markets the INFORUM International System also represents the international financial markets, though in a less detailed way: American interest rates as indicators for international capital market conditions have a weighty influence on German interest rates and this means once again on the German goods markets.

Here PANTA RHEI has been extended by taking into account the influences of the R&D stock on factor demand and final demand. Furthermore the R&D expenditures of the firms were endogenized. Data on sectoral R&D expenditures used in the enlarged model partly were obtained from the Fraunhofer Institute for Applied Systems Analysis so that meaningful R&D capital stock variables could be constructed for most sectors. The construction of the stock variable assumes a 15% depreciation rate. The R&D capital stock variable was found to be significant in the aggregate consumption variable, in several sectoral export functions – in line with expectations especially in the automotive, the investment goods and the electrical engineering sectors – and in sectoral labour demand functions; as regards the latter the sign in most equations was negative and significant, R&D expenditures of a sector are depending from its sales and profits (see appendix).

4.3. The Simulation Results

With this enlarged system we ran simulations from 1999 to 2010. The carbon tax rate is 20 DM per t of CO_2 in 1999 and rises linearly up to 230 DM in 2010. The following tables give some simulation results for the standard case of a carbon tax with full compensation of social security costs and three further simulations, in which the tax revenue is split into a reduction of social security costs and a 5%, 10% and a 15% usage for a support of R&D expenditures of the firms. This means, that the R&D support of the government is distributed to the firms relative to their private R&D activity.

In table 12 the results are given for the year 2010 for GDP and its components, employment and CO_2 emissions as deviations from the bau (business-as-usual)-scenario. In the standard case of a carbon tax with full compensation of social security costs we find a 4.2 % rise of employment hours (or 1m persons additionally employed) and a reduction of CO_2 emissions of 12.9%, which means, that the target of a 25% reduction in relation to the historic value of the year 1990 is reached. This bright picture of a pure ecological tax reform is clouded by the fact that GDP reduces by 3.8%. The latter means that the average growth rate for the next decade would be about 0.3 % lower than in the bau scenario. Since growing employment reduces expenditures of the social security system and raises its revenues and since the government also has a reduction of labour costs, the government has a rise in the financial surplus of 92.6 billion DM.

Tab. 12: Simulation Results for GDP, Employment and CO_2 Emissions
Deviations from the bau* scenario in percent

	Standard Case	Innovation-augmented policy		
		5 %	10%	15%
Gdp	-3.8	-2.4	-0.9	0.5
Consumption	-4.4	-2.8	-1.2	0.4
Government expenditure	-6.8	-5.1	-3.3	-1.6
Equipment	-5.7	-4.3	-2.6	-0.9
Construction	0.4	1.4	2.4	3.3
Exports	-1.2	0.8	2.7	4.7
Imports	-2.5	-0.2	2.2	4.6
Employment in hours	4.2	3.7	3.3	3.0
CO_2 emissions	-12.9	-11.9	-11.0	-10.0

* bau = business as usual

Schumpeterian Ecological Tax Reform

The Schumpeterian scenarios perform better than in the standard scenario concerning GDP development. Already in the 10% case there is nearly no loss in GDP compared to the bau scenario. The positive effects on employment are a little bit weaker than in the standard case: Employment hours rise by 3.3%, but the number of employed persons is with a figure of + 942 000 jobs much nearer to the standard situation, since higher real wages allow a stronger reduction of working time than in the standard case. The reduction of CO_2 emissions is a bit smaller than in the standard case, since we have more growth. But the loss of CO_2 reduction is smaller than the gain in GDP, because the rise of R&D stocks also induces structural change. The growth impulse gives a better performance of the financial surplus of the government.

Tab. 13: Simulation Results for Wages and Prices

- Deviations from the bau scenario in percent

	Standard Case	Innovation-augmented policy		
		5 %	10%	15%
Labour costs per hour	-11.4	-8.5	-6.0	-3.6
Wages per hour	-4.8	-3.0	-0.7	+1.5
Deflator of consumption	4.0	4.1	4.2	4.3

R&D expenditures raise demand and productivity, so that employment is hardly affected. The corresponding effects on wages and prices are depicted in table 13. In the standard case we have a reduction of labour costs per hour compared to the bau scenario of 11.1%, which comes from a 6.3% reduction of social security costs and a 4.8% reduction of wages. The latter results from the reduction in labour productivity of 8% and the rise of consumption prices of 4.0 %. All these are measures against the bau scenario. For the average growth of the real wage in the standard scenario this means growth of only about 1% per year in the next decade.

In the Schumpeterian scenarios we have nearly the same prices for the year 2010, but very different wages compared with the standard case. In the 15% scenario the wage rate will be 1.5% higher than in the bau scenario, because prices are 4.3% higher and productivity is only 2.5% lower than in the bau scenario. Labour costs per hour fall by about 3.6% against the bau scenario. What do these figures mean for the dynamics of the 15% Schumpeterian case? The real wage rate will in this case rise by 1.6-% p.a. over the next decade.

Tab. 14: Simulation Results for Real Gross Production for Selected Sectors and the Industry Average - Deviations from the Bau Scenario in Percent

	Standard Case	Innovation-augmented policy		
		5 %	10%	15%
Chemicals	-1.4	-1.0	-0.5	-0.0
Machinery	-2.4	+1.4	+4.0	+6.6
Motor vehicles	-2.2	+1.3	+3.1	+4.8
Electrical Machinery	-1.8	+1.1	+3.8	+6.4
Industry average	-3.3	-2.0	-0.6	+0.7

To get an impression of the structural effects, which are combined with the innovation-augmented policy, we look in table 14 at results for real gross production of several sectors and the average of all industries. We have chosen the chemical industry and the most important investment sectors machinery, motor vehicles, electrical machinery. These four sectors produce more than 60% of German exports. In the standard case we have a slight reduction of around 2% of their gross production, which is significantly less than in the average of all industries. Especially the investment industries are strongly positively influenced by the innovation-augmented policy. In the 15% case we have positive deviations from the bau scenario up to more than 6%. Also the industrial average will be positively influenced in this case. The gap between the four industries and the average is bigger in the Schumpeterian scenarios than in the standard scenario.

4.4. Some Caveats

Our analysis is just a first step in the study of an innovation-augmented ecological policy. Further research is needed, since we did not analyze externalities as a source of increasing returns. There is diffusion of knowledge between the industries, which allows a firm to take advantage from R&D stocks that have been built by other firms. Such spillovers can have very different channels (GRILICHES, 1995). Since our model PANTA RHEI has a fully integrated and dynamic input-output system, the approach of using the intermediate demand for the calculation of such externalities (SCHERER, 1982; MEYER-KRAHMER/WESSELS, 1989) can easily be endogenized in the system.

At the moment PANTA RHEI has investment only as a vector of final demand. Its source, the investment by industries, is not part of the model. So the effects of R&D on investment could not be studied sufficiently. The next up-date of PANTA RHEI will close this modelling gap, so that a more efficient analysis of R&D on investment can be incorporated.

This revision of the model structure will also allow the use of the investment bridge matrix as a channel for further spillover effects.

5. Optimal Ecological Tax Reform: Options and Recommendations for an EU-Action Plan

An ecological tax reform should bring a true double dividend, namely reduced emissions and higher employment. For several EU countries with high unemployment problems and high labour taxes the option of an ecological tax reform offers some prospect for a double dividend. This holds, however, only if nuclear energy is not artificially subsidized as is still the case in most OECD countries with nuclear energy – not facing the true cost of nuclear energy generation as energy generators face only limited liability (i.e. pay no full insurance) and as nuclear waste storage costs and risks are only partly covered. Only Sweden decided to introduce a special nuclear fuel levy in 1999. Another caveat concerns the expected output reduction in the context of a pure ecological tax reform. The negative output effect can, however, be avoided if government devotes a certain share of ecological tax revenue to growth-enhancing measures where higher R&D promotion is particularly relevant for EU countries, possibly except for Sweden which already has achieved a relatively high R&D-GDP ratio.

There are strong arguments for raising the R&D ratio in EU-15 since economic globalization has intensified the world wide innovation race and since economic opening up and modernization of post-socialist east European economies undermines the comparative advantage of west European firms in many traditional fields of low and medium technology. Moreover, after the end of the Cold War the US has strongly increased the ratio of civilian R&D expenditures to GDP in the 1990s so that the EU faces the risk of falling behind the US in terms of technological progress and economic growth which, of course, is not only relevant for economic reasons but for political reasons as well.

EU coordination in the field of energy taxation would be useful for various reasons. First, it would help to avoid suboptimally low levels of taxation of emissions; secondly, one would largely avoid artificial incentives for costly intra-EU relocation of industrial plants and the EU could make a stronger case for environmental improvement at the global level. Since several EU countries already have introduced some form of ecological taxation successfully and since there are strong theoretical and empirical arguments in favour of a Schumpeterian ecological tax reform governments of EU member countries, the European Commission and the European Parliament should put the topic on the long term political agenda. While one certainly would like to conduct further research on the topic of a Schumpeterian ecological tax reform in EU countries so that refined national policies and optimal harmonization could be achieved the very strategy of combining higher taxes on emissions, lower labour taxes and higher R&D promotion hardly can be refuted for a sustainable economic policy. Therefore the topic of a Schumpeterian ecological tax reform also should be pushed by EU member countries at various international fora, e.g. in the OECD or at the G-7 meetings.

5.1. Optimal Split of Tax Revenue

The optimal split of tax revenue in the context of a Schumpeterian ecological tax reform depends – from a theoretical point of view – on the marginal tax burden (including social security payments to some extent) and the present divergence between social and private benefits of innovation. Empirical analysis will have to shed light on the divergence between the private and social return on innovation in individual EU countries. To the extent that a Schumpeterian tax reform would help to reduce unemployment rates, social security contribution rates (here for unemployment insurance) could be reduced. For EU countries unemployment expenditures relative to GDP reached 1-4.5 % in the mid-1990s which is in marked contrast to the US with a figure of 0.3.

As regards the overall tax burden of labour taxes one should take into account that social security reforms could raise the share of private social expenditures – already some 16% in the EU leaders, the Netherlands and the UK, as compared to 8% and 35% in Germany and the US in 1995 (see Table) – in overall social security while reducing the burden of public social expenditures. One also should note that international differences for net social expenditure-GDP ratios are considerably lower than gross expenditure-GDP ratios (e.g. comparing the US and some Scandinavian countries, high VAT rates in the latter imply higher gross expenditure-GDP ratios for Scandinavia).

Reducing public social expenditures would help to reduce the overall tax burden on labour and hence to promote employment, it probably would also encourage a wider differentiation of social security coverage. Facing the ageing of societies in EU countries the traditional pay-as-you-go social security system will need to be reformed in a broader way. The ecological tax reform should be used not to postpone structural reforms in the social security system but become a starting point for a broader reform.

Tab. 15: Gross and Net Social Security Expenditures in Selected OECD Countries, 1995, (% of GDP)

	AU	B	Can	DK	Fin	Ger	Irl	I	NL	Nor	SW	UK	US
Gross public social expenditure	17,8	28,8	18,2	32,2	31,9	27,1	19,4	23,7	26,8	27,6	33,0	22,4	15,8
Pension spending (old age & survivors)	4,7	10,3	4,8	7,7	9,1	10,9	4,6	13,6	7,8	6,2	9,0	7,3	6,3
Unemployment	1,3	2,8	1,3	4,6	4,0	1,4	2,7	0,9	3,1	1,1	2,3	0,9	0,3
Gross volunt. Private social expenditure	2,8	0,6	4,5	0,9	1,1	0,9	1,8	1,7	4,4	..	2,1	4,2	7,9
Gross total social expenditure	20,9	29,4	22,7	33,6	33,2	29,6	21,1	25,4	31,9	28,5	35,5	27,0	24,1
Private share in social expenditure**	15,0	2,1	19,6	4,1	3,8	8,4	8,3	6,7	16,0	3,1	6,9	16,8	34,7
Net total social expenditure*	21,6	..	21,2	24,4	25,7	27,7	18,7	22,3	25,0	..	27,0	26,0	24,5

*In percent of GDP at factor cost; **in percent of total social expenditures.

Source: Willem Adema (1999), Net Social Expenditure, in: OECD: Labour Market and Social Policy, Occasional Papers, No. 39, Paris

To the extent that government reforms in social security have already raised the share of private social security expenditures to a critical degree, high ecological tax revenues could indeed be used not only for stimulating R&D which makes sense economically only to the point where private and social benefits from innovation are equalized; rather, employment could be further stimulated by reducing (labour) income taxes. Adequate social security reform in combination with lower income taxes and higher innovation dynamics are almost certain to stimulate growth considerably in Europe and the world economy.

5.2. Towards Higher R&D Expenditures in the EU

The R&D-GDP ratio of EU countries could be nearly doubled – i.e. reach some 4% - in the long run if member countries would adopt a Schumpeterian ecological tax reform as proposed here. It is obvious that simply raising government R&D support would be insufficient since with higher R&D promotion it would be crucial to reduce apparent inefficiencies in innovation policies. However, the creation of Euroland and the induced integration of capital markets in the EU have reinforced market pressure on managers to achieve a competitive return on investment and innovation. This pressure could be helpful in both favouring efficient selections of innovation projects on the one hand, while on the other hand it will stimulate EU-wide (and global) diffusion of innovations. To put it differently, the positive supply-side multiplier effects from increased R&D promotion could be higher in the early 21st century than in the 1980s or early 1990s. An increasing role of capital markets also could be helpful with respect to increasing pressure for efficient specialization in R&D; insufficient R&D specialization was an apparent weakness of EU countries in the 1980s and early 1990s (WELFENS/ ADDISON/ AUDRETSCH/ GRUPP, 1998). Prospects for the creation of new technology-oriented firms in the EU would also be improved by the interplay of higher R&D promotion and a greater role of capital markets, including venture capital markets.

With higher R&D expenditures in Europe – and *a fortiori* higher growth – the EU is bound to attract higher external foreign direct investment flows. Germany and Italy as two countries which have had low per capita inflows in the 1990s could thus possibly catch-up with leading FDI host countries in Western Europe, provided that national government adopts adequate reforms which foster the attractiveness of the respective countries as a location for international investors. If national reforms remain insufficient in Germany, and some other EU countries with low FDI inflows, the regional concentration of external FDI inflows in the Community could be reinforced.

5.3. Potential Complementary Measures

An EU-wide Schumpeterian ecological tax reform would generate limited benefits in terms of higher sustainable growth if society would not invest more in higher capital formation at the school and university level (including private universities where appropriate) as well as training and retraining plus distance learning. Skilled labour is likely to benefit strongly from a Schumpeterian ecological tax reform, unskilled labour would also benefit to the extent that substitution effects in favour of labour inputs, and growth effects, induce a higher aggregate labour demand. With respect to unskilled labour one should, however, note the caveat that a higher R&D capital stock naturally raises labour productivity, which in turn could reduce labour demand slightly, especially with respect to unskilled labour. Special training and retraining efforts thus might be required in particular for unskilled workers and for the long term unemployed, respectively.

As regards efficient reduction of emissions it is appropriate with respect to certain emissions – e.g. CO_2 and SO_2 – also to stimulate global and regional emissions trading. Taking into account the issue of EU competitiveness, cost-minimizing ways of cutting CO_2 are important. As the early US experiences from emissions trading are positive, the EU might also want to pursue this avenue of modern environmental policy. Together with spot and forward electricity markets that will emerge in the context of EU electricity liberalization the creation of markets for emission certificates could boost financial markets in Europe and contribute to major efficiency and welfare gains.

5.4. Recommendation of EU Action Plan

We recommend – on the basis of present research – the following elements of an EU action plan:

- Introducing a Schumpeterian ecological tax reform in each country, leaving to the respective member state the choice of splitting the revenue from ecological taxes between R&D promotion and cutting social security payments/income taxes.
- Adopting common minimum CO_2 and energy taxes in all EU countries.
- Monitoring the environment and the development of national and sectoral R&D stocks, respectively.
- Strengthening private expenditures on social security.
- Encouraging labour market flexibility and regional mobility.
- Stimulating innovation and diffusion across EU member countries, including measures of the European Commission in the context of Article 10 EFRE funds (0.5% of structural funds can be devoted to innovative projects; similar rules apply to ESF/EAGFL funds).
- Extending the Schumpeterian ecological tax reform to Eastern Europe and the Balkans within a European Pact for Innovation, Environment and Growth

5.5. Facing Global Energy Efficiency Issues

By bringing the topic of a Schumpeterian ecological tax reform on the OECD and G-7/8 agenda and by introducing this concept in UN conferences the EU could become an engine of efficiency-guided policy reforms worldwide. There is little doubt that EU countries – along with the US and Japan – have a competitive advantage in developing new products and process innovations so that Europe could strongly benefit from a global move towards a Schumpeterian ecological tax reform. It would be useful if Germany and other leading EU countries could adopt successful models of ecological tax reforms early on.

If EU countries were successfully to adept major policy innovations this would not only help to cope with major economic and ecological problems. This also would demonstrate that a club of diverse countries could join forces to pursue new ways of achieving prosperity and stability. There is, however, some risk that a Schumpeterian ecological tax reform could undermine reforms in the field of social security in the EU, although reforms are quite urgent in many greying economies where the share of retired persons to the number of people working is declining dramatically. Rising ecological tax revenues should be devoted to reducing social security contributions only to some extent.

Given the high-expected growth of world output in the 21st century, strengthening innovation in combination with environmental improvement is crucial for the global community. If the EU were to establish leadership in the field of Schumpeterian ecological reform it would not only remedy serious European problems but it also would prove an ability to influence the global agenda in a meaningful way. With respect to CO_2 emissions India and China will play a crucial role in the 21st century for global warming so that transfer of Schumpeterian policy concepts and technology transfer via high foreign direct investment flows could play a crucial role in the future. Raising energy efficiency is a topic of common interest also for the EU (plus the US and Japan) and Russia so that new policy initiatives and joint international research projects could be developed.

A Schumpeterian ecological tax reform in the EU could reinforce the economic dynamics of Europe and contribute to global environmental improvement. New Schumpeterian dynamics in the EU and proven environmental global progress would help to reinforce the role of Europe worldwide.

6. Conclusions

Germany's energy policy has introduced new elements in the form of phasing-out nuclear energy and of introducing an ecological tax reform. At the core of energy policies in the 1990s is the electricity and gas market. With the share of nuclear energy declining in the long run the import of gas and gas-fired power plants are likely to become more important – creating new opportunities for importing gas from Russia which, however, represents some supply risk due to political uncertainties and economic instabilities (WELFENS, 1999c). Gas imports and electricity imports, respectively, will increase in the long run. There are also prospects for rising German/EU electricity imports from transforming economies in eastern Europe where governments increasingly are aware of the need to combine modernization of the energy sector with requirements of environmental protection; but with respect to individual countries of the Visegrad group there are considerable differences (WELFENS/ YARROW, 1997; JASINSKI/ SKOCZNY, 1996a and 1996b; JASINSKI/ PFAFFENBERGER, 1999; WELFENS/ GRAACK/ GRINBERG/ YARROW, 1999).

The gradual liberalization of the EU electricity market and full German liberalization of this market will contribute to rising electricity trade (WELFENS, 1999b). Germany is likely to rely on higher electricity imports on the one hand, while on the other hand, there will be stronger incentives to switch to more energy-efficient power generation, including increasingly combined heat and power generation; the use of renewable energy sources also could play a slightly increased role, although intensified price competition in the context of a liberalized EU electricity market could undermine earlier prospects for a rapid expansion for the share of renewables.

As practically the whole stock of fossil power plants will have to be replaced in the period 2000-2030 there is a high potential for combined heat and power plants, and phasing-out nuclear energy could create additional investment needs (PFAFFENBERGER, 1999). Facing falling electricity prices the political will to encourage expansion of combined heat and power plants will remain important for reaching higher energy efficiency in Germany. Since the CO_2 emissions (per kWh) from gas fired power plants is only half that of lignite we can expect policy will have to restrict the role of lignite if CO_2 emissions reduction targets are to be achieved. Taking into account the goal of phasing out nuclear energy one has to anticipate a strongly increasing German gas demand, which could contribute to higher gas prices in Europe and worldwide. A reduction of CO_2 emissions in the electricity sector is possible to some degree with a higher share of natural gas. Phasing out nuclear energy power plants and replacing them by less capital intensive natural gas and coal power plants could contribute to temporary job losses.

A greater role for gas, in turn, will mean increasing price risks with respect to changes in the natural gas price. One should also note that Euroland's current account position would suffer from rising German gas imports. The additional gas imports mainly could come from Russia but also from other countries. Germany and the Community could deliberately use rising gas imports from Russia as a means to stabilize the Russian transformation process. However, if Russia is not adopting adequate internal reform policies – with a strong focus on the rule of law, budgetary and tax reforms plus the creation of a credible and efficient banking system – rising Russian exports of natural gas could mainly result in increasing capital flight and rising Mafia activities undermining the goal of sustainable economic transformation.

The ecological tax reform adopted and envisaged in Germany is only partly consistent and adequate with respect to reducing emissions and raising employment. Ecological taxes are not explicitly differentiated according to the CO_2 content; the German approach adopted by the red-green coalition in late 1998 thus is in contrast to that in the Netherlands, Denmark and Sweden (SCHLEGELMILCH, 1999). The main problem of a pure ecological tax reform envisaged in Germany – raising energy taxes and reducing social security contributions – is that this strategy lacks a policy element which would offset the growth-reducing effects of the tax reform. Based on a broader theoretical framework which emphasised the need of optimal factor allocation which in turn requires internalization of positive and negative external effects plus reduction of tax-induced distortions a Schumpeterian ecological tax reform is proposed (MEYER/WELFENS, 1999): Part of ecological tax revenues would be used to raise expenditures on research and development – internalizing positive external effects - so that growth is reinforced; about 1/10 of the tax revenue has to be devoted to higher R&D expenditures if a negative growth effect from ecological taxes is to be avoided. From a theoretical point of view raising energy/ CO_2 taxes should contribute to internalizing negative external effects. The simulations show that adequate CO_2 taxation can help Germany to achieve the Kyoto emission reduction goals. While the parallel reduction of social security contributions together with the simulated output path suggests that about 1 million new jobs could be created by 2010 – going along with efficiency gains due to lower labour tax distortions – it is obvious that a Schumpeterian ecological tax reform can be only part of a broader policy strategy to regain full employment and to strengthen international competitiveness in Germany and the EU, respectively.

We recommend that the German government adjusts its ecological tax reform in at least three ways: a) switching from an electricity tax towards an explicit CO_2 tax, b) introducing an element of promoting R&D strongly in order to reinforce economic growth and to encourage emission-saving product and process innovations, c) seeking a broader EU consensus on a Schumpeterian ecological tax reform. Moreover, government should base tax rates chosen on adequate scientific analysis so that tax rate is neither excessive nor too low to achieve an internalization of external effects. It is obvious that the scientific basis of modern economic policy in Germany could be improved, the improved efficiency and effectiveness of scientifically based strategies could help government to avoid costly policy pitfalls and reinforce the legitimacy of its reform approach; finally, a consistent and efficient tax reform approach also will facilitate the shoring up broader EU and G-7 support for similar reform strategies – this in turn would make international Cupertino easier. Comprehensive reforms (beyond the ecological tax reform) of the social security system and of labour markets are necessary in Germany (ADDISON/WELFENS, 2000).

Taking into account EU growth prospects of 2-2.5% p.a. in the period 2000-2020 Community GDP will increase by more than 50%, while traffic is likely to more than double within the same period so that emission levels and the resource use will increase strongly. Improving energy efficiency therefore will remain a crucial challenge for the EU, particularly in the fields of electricity and heat generation as well as in traffic. In this context an innovation-oriented ecological tax reform could be applied in almost all EU countries yielding high long-term benefits. Except for Sweden where the R&D-GDP ratio had reached 3.9% in 1997 already this ratio was below 3% in all other EU countries – with Italy achieving roughly 1% only which seems to be insufficient for a high wage country with only modest growth but high unemployment. Without further investigation it is unclear how strongly the R&D-GDP ratios should increase in individual EU countries and in the Community respectively; the EU has spent 1.83% of GDP on R&D in 1997 which seems to be inadequate relative to Japan (2.92) and the US (2.79%), especially since the latter has strongly raised its civilian share of R&D in overall R&D expenditures. It remains a challenge for future research to determine more exactly – and on the basis of empirical investigation – the opportunities for an EU-wide ecological tax reform.

From a theoretical perspective one may argue that all countries with a high conventional tax burden should raise ecological taxes in order to internalize negative external effects – the broad tax-GDP ratio (sum of direct taxes, indirect taxes and social security contributions as % of GDP: EUROPEAN COMMISSION, 1999, p.19) in all Euroland countries exceeded 40% in 1998 except for Ireland (32.3), Spain (36.9) and Portugal (36.5). It is obvious that an innovation-oriented ecological tax reform could be particularly useful for countries with high unemployment rates, and this holds the more the higher the potential growth prospects for R&D-intensive products in output and exports are. However, one should study more carefully the macroeconomic and sectoral aspects of an EU-wide innovation-oriented ecological tax reform, e.g. based on an adequately extended version of the PANTA-RHEI model with its unique approach of combining input-output analysis, emission modelling, use of R&D capital stocks and standard macroeconomic accounting.

Increasing government R&D promotion strongly within a Schumpeterian oriented ecological tax reform – as proposed here - raises several issues, including that of raising the efficiency of government R&D policies in EU countries which are known to suffer from various problems in comparison to the US (WELFENS/ AUDRETSCH/ ADDISON/ GRUPP, 1998). EU countries will have to spend more on R&D and in some cases should specialize more strongly in the innovation race; moreover, new technology fields should be promoted adequately by government R&D policies which mainly require improved conditions for start-up companies and venture capital financing, respectively. Reforming higher education in a way which makes the university system more responsive to the needs of a high technology (and the information services) society also is important. As regards Germany and some other EU countries one could also recommend improving conditions for foreign direct investment inflows since multinational companies′ investment plays a core role in raising R&D intensity. Moreover, strengthening the role of capital markets by a broad social security reform which reinforces voluntary savings for retirement is advisable since broader capital markets would put adequate pressure on managers to adopt adequate innovation strategies which reinforce long-term profitability of the firm.

Perspectives

A Schumpeterian ecological tax reform could not only bring major benefits for Germany but for the whole EU (it also would be applicable in the US, Canada and Japan). Since in 1997 per capita CO_2 emissions of Belgium (12 tons) and the Netherlands (11.8 t) were somewhat higher than in Germany (10.8 t) the benefits of a Schumpeterian ecological tax reform in these two countries could be even higher than in Germany. With British, Italian and French per capita emissions being slightly lower – that 9.4, 7.4 and 6.2 percent, respectively – a Schumpeterian ecological tax reform would also generate considerable benefits in the UK, Italy and France; the latter, of course, faces special problems from its large nuclear power industry. It would be useful to adopt an EU harmonization approach in the field of (Schumpeterian) ecological tax reform, the two main reasons being that some minimum harmonization would avoid distortionary effects on intra-Community trade and foreign direct investment on the one hand, on the other hand taxation of primary energy inputs according to the respective CO_2 intensity hardly is possible without an EU-wide approach. Germany′s ecological tax reform should be modified adequately which means both adopting an explicit focus on CO_2 emissions of primary energy and splitting the ecological tax revenue in a way which would not only reduce labour costs but also raise R&D promotion and R&D expenditures, respectively.

The EU should adopt a Schumpeterian ecological tax reform project, which would generate new jobs, strengthen EU competitiveness and stimulate economic growth. Facing the New World of economic globalization an adequate EU action program could considerably contribute to modernizing the EU and to successfully cope with global warming.

Appendix

Tab. A1: Electricity Prices for EU Industry on 1 January 1999

Country	Industrial users[3]	Electricity prices[1] (Euro / 100kWh)
Norway – National	La	5.09
	Li	2.20
Finland – National	La	5.49
	Li	2.68
Sweden – National	La	6.20
	Li	2.56
Denmark – National	La	5.31
	Li	4.21
Greece – Athens	La	8.61
	Li	4.03
France – Paris	La	8.91
	Li	4.17
Netherlands – Rotterdam	La	10.09
	Li	4.54
Spain – Madrid	La	9.77
	Li	4.86
Portugal – Lisbon	La	10.66
	Li	4.34
United Kingdom – London	La	9.83
	Le[2]	5.94
Ireland – Dublin	La	12.78
	Li	4.85
Luxembourg – National	La	13.66
	Li	4.34
Italy	La	14.41
	Li	4.16
Belgium – National	La	14.64
	Li	4.15
Germany – Düsseldorf	La	13.57
	Li	5.55
Austria – Oberösterreich/Tyrol/Vienna	La	14.90
	Li	5.11

[1] taxes excluded

[2] no dates for li provided

[3] standard industrial users:

Standard consumer	Annual consumption (kWh)	Maximum demand (kWh)	annual utilisation (in hours)
La	30 000	30	1 000
Le	2 000 000	500	4 000
Li	70 000 000	10 000	7 000

Source: Eurostat (1999): Statistics in focus, Environment and Energy, Theme 8 – 3/1999.

Tab. A2: Gas Prices for EU Industry on 1 January 1999

Country	**Industrial users**[2,3]	**natural gas prices**[1] (Euro / GJ)
Finland – National	L2	4.24
	L5	2.49
Denmark – Copenhagen	L1	4.70
	L4-2	2.12
United Kingdom – London	L1	4.21
	L4-1	3.02
Germany – Düsseldorf	L1	5.50
	L5	1.97
Belgium – National	L1	5.59
	L4-2	2.00
Netherlands – Rotterdam	L1	4.93
	L3-2	3.09
France – Paris	L1	6.14
	L4-2	2.40
Luxembourg – City of Luxembourg	L1	5.25
	L4-2	3.72
Sweden – Malmö	L1	5.89
	L3-1	3.37
Spain – Madrid	L1	6.54
	L5	2.53
Italy – Milan	L1	7.24
	L4-2	2.76
Ireland – Dublin	L1	7.25
	L3-2	3.09
Austria – Vienna	L1	7.60
	L4-1	3.71

[1] taxes excluded

[2] lowest and highest standard consumer for each country

[3] standard industrial users:

Standard Consumer	Annual consumption	Modulation
L1	418.60 GJ (i.e. 116 300 kWh)	no load factor laid down
L2	4 186 GJ (i.e. 163 000 kWh)	200 days
L3-1	41 860 GJ (i.e. 11.63 GWh)	200 days 1 600 hours
13-2	41 86 GJ (i.e. 11.63 GWh)	250 days 4 000 hours
14-1	418 600 GJ (i.e. 116.30 GWh)	250 days 4 000 hours
14-2	418 600 GJ (i.e. 116.30 GWh)	330 days 8 000 hours
15	4 186 000 GJ (i.e. 1163.00 GWh)	330 days 8 000 hours

Source: Eurostat (1999): Statistics in focus, Environment and Energy, Theme 8 – 5/1999.

Tab. A3: Export Functions. Elasticities, t-Statistics (in brackets), R^2 and D.W., Germany

Coefficients (DW)Sector	Abs	$krfe_i$	Peg_i/fpe_i	fdm_I	Ttrend	D91FF	R^2	DW
15	-0.84 (-0.91)	0.26 (2.75)	-0.44 (-2.62)	0.54 (2.50)	-	-0.88 (-3.47)	0.98	2.14
21	0.69 (17.89)	0.87 3.41	-	0.58 (4.49)	-0.04 (-4.01)	-	0.88	1.57
23	0.21 (1.62)	0.35 (1.13)	-0.28 (-2.28)	0.53 (1.48)	-	-0.12 (-2.50)	0.96	2.08
26	-0.06 (-0.79)	0.86 (1.91)	-0.20 (-1.99)	0.18 (0.49)	-	-	0.98	1.79
28	-	0.12 (4.16)	-0.18 (-1.78)	0.60 (86.34)	-	-0.09 (2.79)	0.95	2.22

Abs: Intercept
$krfe_i$: R&D capital stocks of sector i
peg_i: German Export Price of good i
fpe_i: World market price index of good i
fdm_i: World import demand of good i
Ttrend: Time trend
D91FF: Dummy variable (0 until 1990, 1 from 1991)

Tab. A4: Investment in Construction Functions. Elasticities, t-Statistics (in brackets), R^2 and Durbin Watson Coefficients (DW), Germany

Sector	Abs	KRFE	RUMLR	$RUMLR_{t-1}$	D91FF	R^2	DW
19		0.36 (5.77)	-0.03 (-1.45)		0.03 (5.51)	0.97	2.03
20	0.35 (2.55)	0.31 (4.55)	-0.03 (-2.42)		0.02 (4.96)	0.96	1.34
26	-0.77 (-4.28)	0.53 (9.60)	-0.06 (-3.91)		0.02 (5.49)	0.97	1.57
41	0.93 (15.96)	0.08 (1.52)		-0.04 (-5.57)	0.01 (9.02)	0.96	2.10
42	0.78 (10.82)	0.18 (3.06)		-0.03 (-4.00)	0.02 (10.30)	0.97	1.57

Abs: Intercept

KRFE: R&D capital stock of all sectors

RUMLR: Yield of government bonds minus inflation rate

D91FF: Dummy variable (0 until 1990, 1 in 1991, 0 from 1992)

Tab. A5: Investment in Equipment Functions. Elasticities, t-Statistics (in brackets), R^2 and Durbin Watson Coefficients (DW), Germany

Sector	KRFE	GG/PG	RUMLR	D91	R^2	DW
26	0.64 (1.77)	0.49 (0.34)	-0.05 (-1.08)	0.00 (-3.87)	0.98	2.08

KRFE: R&D capital stock of all sectors

GG/PG: Profit of all sectors in prices of 1991

Tab. A6: Macro Consumption Function. Elasticities, t-Statistics (in brackets), R^2 and Durbin Watson Coefficients (DW), Germany

Abs	YVH	KRFE	RKONTR	D91FF	R^2	DW
0.14 (6.38)	0.79 (12.74)	0.08 (4.00)	-0.02 (-3.81)	0.01 (4.96)	0.99	2.42

YVH: Disposable Income of private households in prices of 1991

RKONTR: Interest rate on bank loans minus inflation rate

Tab. A7: Labour Demand Functions. Elasticities, t-Statistics (in brackets), R^2 and Durbin Watson Coefficients (DW), Germany

Sector	Abs	$krfe_i$	xg_I	wgr_i	D91FF	R^2	DW
3	0.28 (4.36)	0.28 (-4.19)	0.98 (10.98)	-	-	0.90	1.34
6	0.38 (4.65)	-0.09 (-1.25)	1.04 (7.24)	-	-	0.80	1.48
11	0.68 (9.36)	-0.12 (-3.28)	0.95 (7.49)	-0.50 (-2.37)	0.00 (1.32)	0.96	1.26
12	1.05 (16.48)	-0.17 (-10.34)	0.82 (6.73)	-0.69 (-2.55)	-	0.94	1.55
13	0.94 (28.10)	-0.05 (-2.32)	0.98 (33.47)	-0.99 (-16.99)	-	0.99	2.36
15	1.59 (11.03)	-0.15 (-3.66)	0.84 (8.32)	-0.77 (-5.80)	0.00 (1.53)	0.95	1.28
19	0.89 (25.91)	-0.30 (-4.72)	0.74 (11.25)	-0.71 (-9.34)	0.00 (6.21)	0.95	1.57
20	1.33 (11.97)	0.19 (-1.45)	0.32 (2.62)	-1.00 (-6.05)	0.01 (2.40)	0.98	1.35
21	0.90 (10.16)	-0.11 (-1.11)	0.71 (6.41)	-0.72 (-4.30)	0.00 (5.79)	0.92	1.30
22	0.81	-0.13	0.34	-	-	0.58	1.17
23	0.81 (10.38)	-0.32 (-3.33)	0.72 (6.92)	-0.38 (-2.49)	0.00 ((5.10)	0.88	1.86
24	1.36 (18.83)	-0.61 (-9.87)	0.76 (11.21)	-0.91 (-12.83)	0.03 (18.52)	0.99	1.85
26	0.88 (22.77)	-0.10 (-0.72)	0.92 (6.26)	-0.94 (-10.02)	0.00 (1.87)	0.94	1.45

Tab. A7: Labour Demand Functions. Elasticities, t-Statistics (in brackets), R^2 and Durbin Watson Coefficients (DW), Germany (cont.)

27	1.01	-0.60	0.68	-0.47	0.01	0.90	1.55
	(8.27)	(-1.52)	(5.55)	(-1.41)	(4.21)		
28	0.91	-0.05	0.80	-0.75	0.00	0.96	2.13
	(27.21)	(-3.04)	(14.17)	(-8.91)	(3.31)		
31	1.28	-0.03	0.46	-1.10	0.00	0.97	1.30
	(19.27)	(-1.52)	(5.22)	(-10.27)	(5.08)		
32	0.90	-0.14	0.89	-0.69	0.01	0.92	1.35
	(8.86)	(-3.66)	(5.96)	(-10.77)	(5.49)		
33	0.97	-0.12	0.72	-0.71	0.01	0.90	1.31
	(17.31)	(-5.04)	(6.91)	(-7.06)	(5.22)		
34	1.15	-0.02	0.45	-0.76	0.01	0.92	2.05
	(18.07)	(-1.41)	(4.71)	(-7.29	(5.76)		
35	1.33	-0.26	0.89	-1.35	0.00	0.98	1.07
	(8.32)	(-2.09)	(4.18)	(-8.72)	(1.26)		
36	0.95	-0.36	0.57	-0.68	-	0.97	1.22
	(10.14)	(-3.53)	(3.50)	(-3.79)			
37	1.11	-0.18	0.88	-1.24	-	0.99	1.11
	(15.64)	(-2.39)	(7.05)	(-12.63)			
39	1.44	-0.09	0.20	-0.96	0.01	0.99	1.38
	(20.28)	(-7.05)	(1.92)	(-30.15)	(11.06)		
40	1.77	-0.43	-	-1.09	-0.02	0.95	1.96
	(11.59)	(-6.23)		(-10.02)	(8.23)		

xg_i: Gross productions in sector i in Prices of 1991

wgr_i: real wage rate in sector i

Tab. A8: Negative External Effects of Energy-Related Emissions: International Comparison

Country	Rio emission target reduction (basis 1991)*: % p.a.	Share in worldwide energy-related CO_2 (%)	GDP per capita 1992 (in $)	Energy intensity of industry in GJ/1000DM
	EREDV	SHARE	GDP	E/GDP
Luxembourg	0,00	0,05	31343	5,00
Greece	3,13	0,34	7686	3,54
Switzerland	-0,28	0,20	35606	3,51
Portugal	2,76	0,19	8534	3,50
Australia	-1,21	1,24	16715	3,46
Spain	2,50	1,02	14697	3,27
Netherlands	-1,14	0,74	21130	3,17
France	0,40	1,74	23149	3,13
Germany	-1,73	4,60	24157	3,07**
Ireland	2,50	0,16	14484	2,98
Austria	-2,5	0,27	23725	2,97
Japan	0,75	4,97	29387	2,96
Belgium	-1,53	0,52	21935	2,95
Canada	0,25	2,02	20600	2,95
Italy	-0,18	1,92	21177	2,94
Finland	-1,04	0,25	21756	2,84
USA	0,25	22,70	2332	2,81
Norway	0,39	0,14	26331	2,80
Denmark	-2,03	0,25	27626	2,78
New Zealand	-0,48	0,12	12003	2,77
Sweden	0,24	0,25	28291	2,75

* National targets recalculated for 1991 basis; ** West Germany

Source: (ROTHFELS, 1999)

Tab. A9: Electricity Generation in Germany According to Energy Sources in TWh

	1991	1992	1993	1994	1995	1996	1997	1998
Hard Coal	149.8	141.9	146.2	144.6	147.1	152.7	143.1	151.5
Lignite	158.4	154.5	147.5	146.1	142.6	144.3	141.7	140
Heating oil	13.6	12	8.9	8.8	7.8	6.9	5.9	5.5
Natural gas	36.3	33	32.8	36.1	41.1	45.6	48	51.5
Other fuels	15.4	15.8	15.3	17.5	18	17.5	19.8	21.5
Nuclear energy	147.4	158.8	153.5	151.2	154.1	161.6	170.3	161.5
Hydro	18.5	21.1	21.5	22.5	24.2	21.7	20.9	20.5
Total	539.4	537.1	525.7	526.8	534.9	550.3	549.7	552

Source: BMWi, Energiedaten '99

Tab. A10: Electricity Generation in Germany According to Energy Sources in %

%	1991	1992	1993	1994	1995	1996	1997	1998
Hard Coal	27.77	26.42	27.81	27.45	27.50	27.75	26.03	27.45
Lignite	29.37	28.77	28.06	27.73	26.66	26.22	25.78	25.36
Heating oil	2.52	2.23	1.69	1.67	1.46	1.25	1.07	1.00
Natural gas	6.73	6.14	6.24	6.85	7.68	8.29	8.73	9.33
Other fuels	2.86	2.94	2.91	3.32	3.37	3.18	3.60	3.89
Nuclear energy	27.33	29.57	29.20	28.70	28.81	29.37	30.98	29.26
Hydro	3.43	3.93	4.09	4.27	4.52	3.94	3.80	3.71
Total	100	100	100	100	100	100	100	100

Source: BMWi, Energiedaten '99

Tab. A11: Energy Security and Germany's External Dimension of Energy Supply

Imports of Primary Energy according to Countries 1997

Crude Oil			Natural Gas		
Countries	1000 t	%	Countries	Billion kWh	%
Near East	11850.0	12.0	Netherlands	231.1	29.2
Africa	20693.0	20.9	Norway	209.7	26.5
Venezuela	2443.0	2.5	Former SU	319.8	40.4
Former SU	25483.0	25.7	Others	30.6	3.9
Norway	21812.0	22.0			
Great Britain	16662.0	16.8			
Total	98993.0	100.0	Total	791.2	100.0

Source: BMWi. Energiedaten '99

Tab. A12: Energy Imports and Exports, 1997

	Crude Oil	Crude Oil-Products	Natural Gas	Electricity
Imports	98.993 Million t	49.302 Million. t	791.2 Billion kWh	38.0 TWh
Exports		12.665 Million t	38.5 Billion kWh	40.4 TWh

Source: BMWi. Energiedaten '99

Tab. A13: Real Value of Primary Energy Imports and Exports of Germany 1997

Million DM	Crude Oil	Fuels. Lubricants. Natural Gas
Imports	25.199.7	27.996.8
Exports	221.0	8.083.8

Source: Statistisches Jahrbuch für die Bundesrepublik Deutschland. 1998

Fig. A1: Natural Gas Consumption, 1970 to 1997, UK

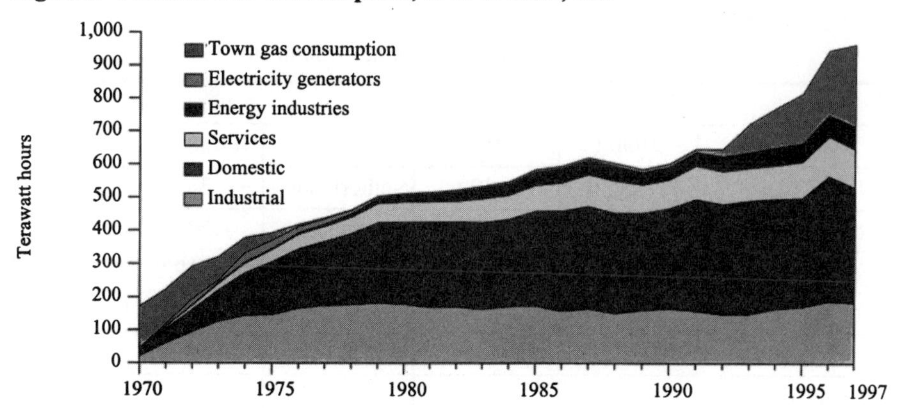

Tab. A14: Natural Gas (Includes Substitute Natural Gas and Colliery Methane), UK, in TWh

	1970	1980	1990	1994	1995	1996	1997
Electricity generators	1.9	4.0	8.2	118.0	149.4	195.0	248.3
Energy Industries	1.2	19.1	39.2	54.9	57.0	65.5	67.9
Industry	20.8	177.5	165.0	164.5	170.2	186.9	182.3
Domestic	18.4	246.8	300.4	329.7	326.0	375.8	345.5
Services	3.9	59.4	86.8	101.7	110.5	120.6	117.6
Total	46.2	506.8	599.6	768.8	813.2	943.8	961.6

Fig. A2: Coal Production, 1970 to 1997, UK

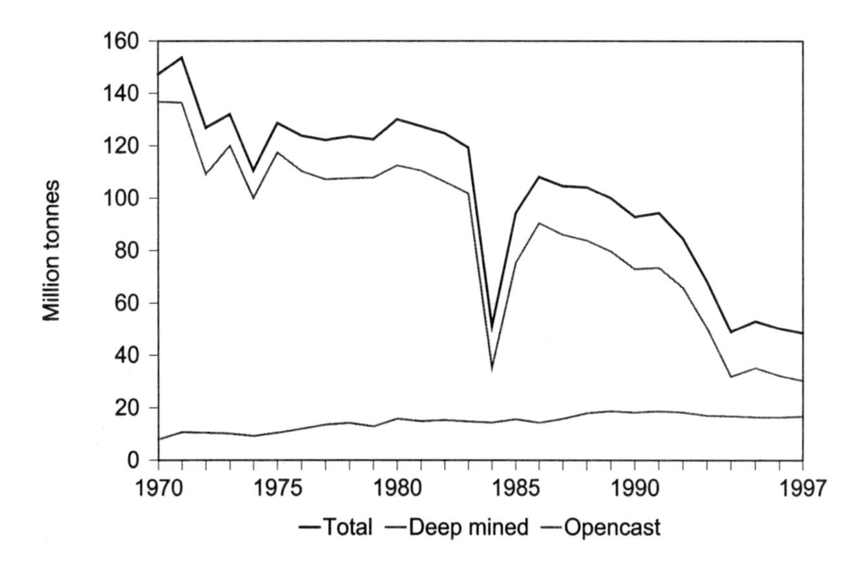

Tab. A15: Coal Mining, Million Tonnes, UK

	1970	1980	1990	1995	1996	1997
Deep mined	136.7	112.4	72.9	35.1	32.2	30.3
Opencast	7.9	15.8	18.1	16.4	16.3	16.7
Total (including slurry)	147.2	130.1	92.8	53.0	50.2	48.5

Fig. A3: Coal Consumption, 1970 to 1997, UK

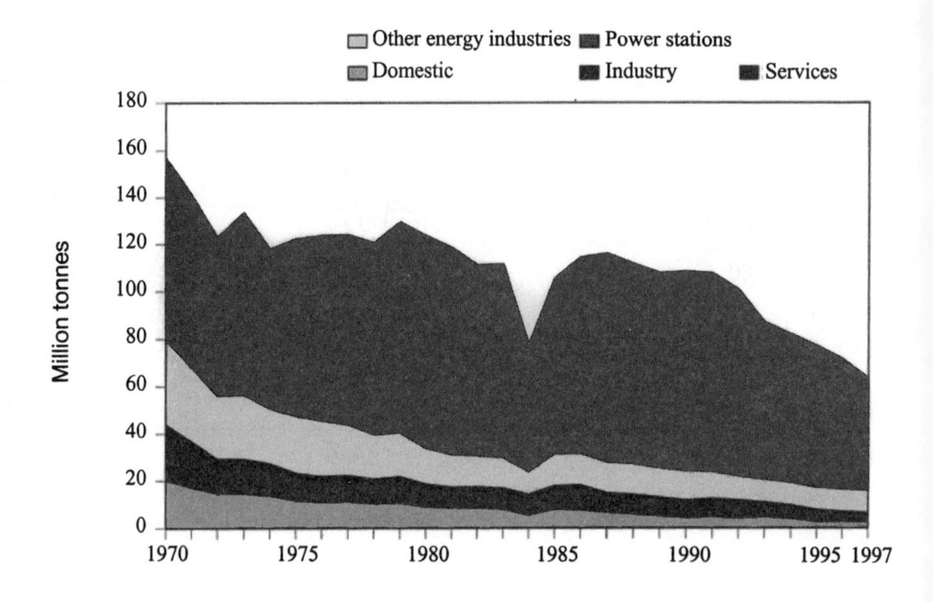

Tab. A16: Coal Consumption by Sectors, in Million Tonnes, UK

	1970	1980	1990	1995	1996	1997
Power stations	77.2	89.6	84.0	59.6	54.9	47.1
Domestic	20.2	8.9	4.2	2.7	2.7	2.6
Industry	19.6	7.9	6.3	4.5	3.6	3.2
Services	4.2	1.8	1.2	0.5	0.6	0.6
Other energy industries	35.7	15.3	12.5	9.7	9.6	9.6
Total consumption	156.9	123.5	108.3	76.9	71.4	63.1

Fig. A4: Gross Electricity Supplied by Nuclear Generation, 1970 to 1997, UK

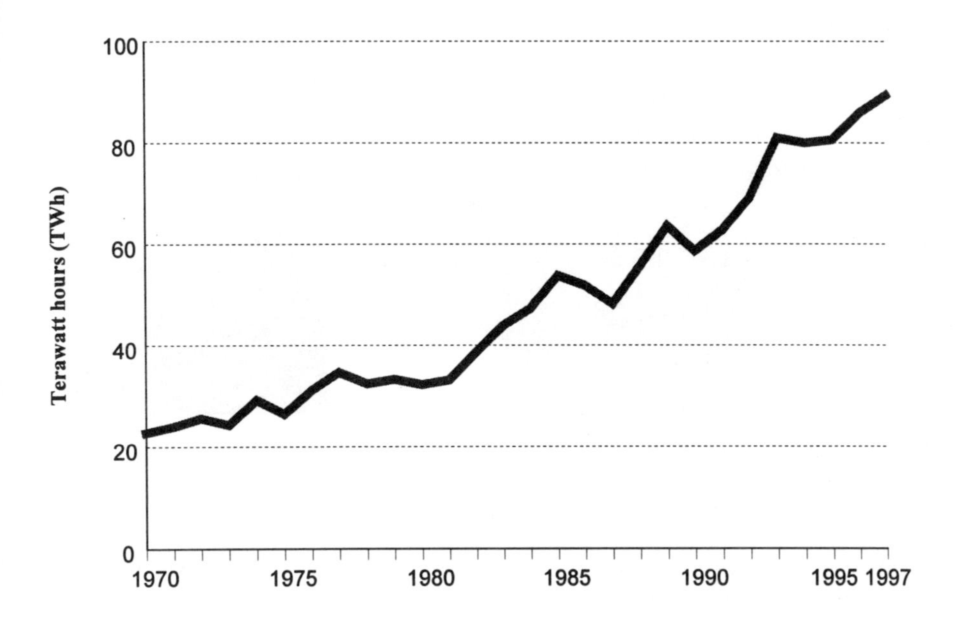

Tab. A17: Gross Electricity Supplied by Nuclear Generation, 1970 to 1997, UK

	1990	1995	1997
Electricity supplied (gross)	59 TWh	81 TWh	89 TWh
% of electricity generation	21%	27%	28%
Employment	42,500	26,000	30,500
Turnover[1]	4.5bn	4.9bn	4.4bn
Gross value added	2.2bn	3.2bn	3.2bn
% contribution to GDP	0.46%	0.53%	0.45%
Exports	300m	500m+	600m

1. Includes revenue from the nuclear premium.
All figures are in current prices.

Fig. A5: Renewable Energy Sources, UK

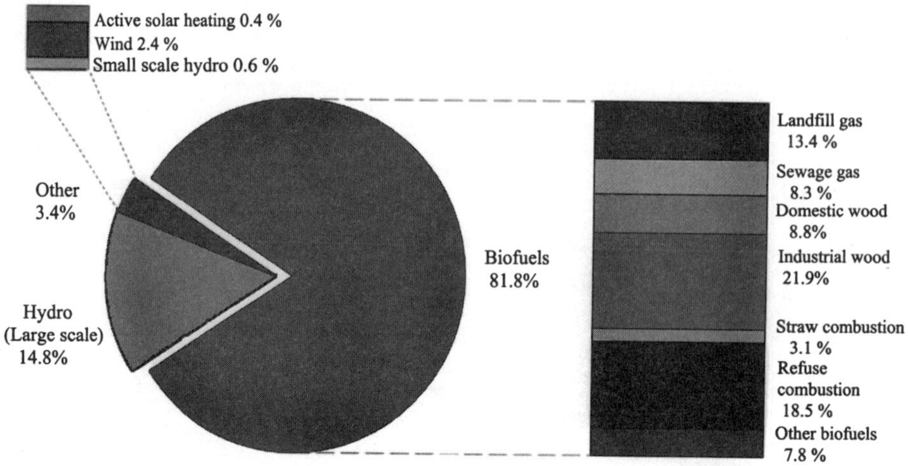

Total renewables used = 2.31 million tonnes of oil equivalent

Tab. A18: Total Use of Renewable (Thousand Tonnes of Oil Equivalent), UK

	1990	1995	1996	1997
Active solar heating	6.4	8.2	8.6	9.0
Onshore wind	0.8	33.6	41.8	57.2
Hydro	447.7	416.0	289.0	354.8
Landfill gas	79.8	205.8	251.5	308.7
Sewage sludge digestion	138.2	173.5	190.2	191.5
Wood	174.1	702.3	709.7	710.3
Straw	71.7	71.7	71.7	71.7
Municipal solid waste	160.0	357.6	368.7	427.0
Other biofuels	80.6	180.0	175.8	182.0
Total	1,159.3	2,148.7	2,107.0	2,312.2

Fig. A6: Fuel Use for Electricity Generation, 1970 to 1997, UK

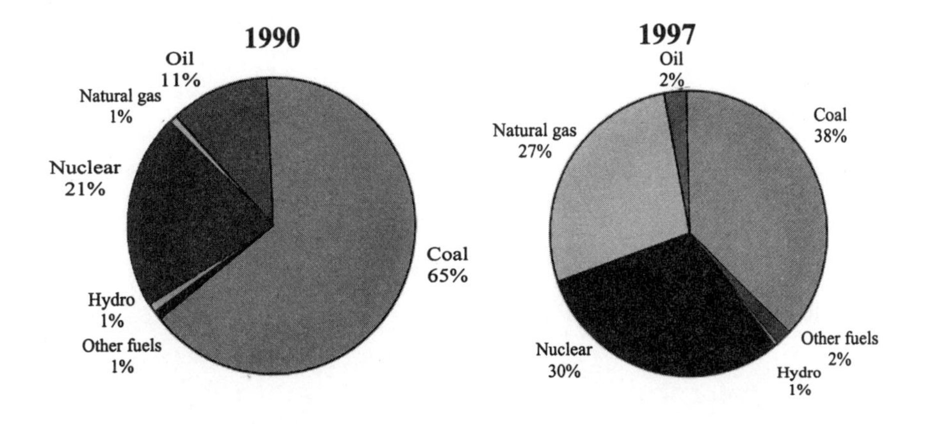

Tab. A19: Fuel Use for Electricity Generation, 1970 to 1997 (Million t of Oil Equivalent), UK

	1970	1980	1990	1995	1996	1997
Coal	43.1	51.0	49.8	36.1	33.0	28.6
Oil	13.3	7.7	8.4	3.6	3.5	1.9
Gas	0.1	0.4	0.6	12.5	16.4	20.9
Nuclear	7.0	9.9	16.3	21.3	22.2	23.0
Hydro	0.4	0.4	0.4	0.4	0.3	0.3
Other fuels	-	0.1	0.8	1.1	1.2	1.4
Total	63.9	69.5	76.3	75.1	76.6	76.1

Fig. A7: Natural Gas Consumption, 1970 to 1997, UK

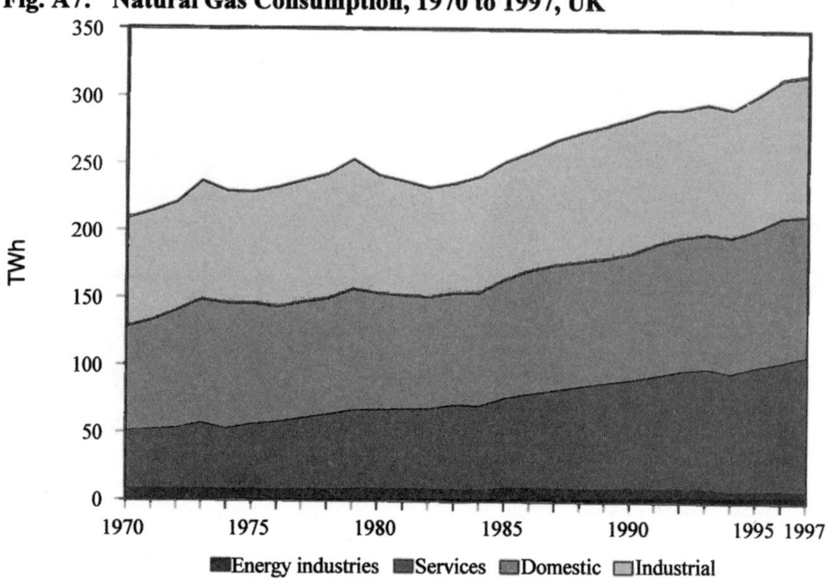

Tab. A20: Electricity Consumption by Sectors, 1970 - 1990 (TWh), and UK

	1970	1980	1990	1995	1996	1997
Energy industries	8.2	8.6	10.0	8.3	8.6	8.2
Industry	81.1	88.6	100.6	99.9	103.1	104.7
Domestic	77.0	86.1	93.8	102.2	107.5	104.5
Services	42.4	58.4	80.0	91.8	95.0	100.1
Total	208.7	241.7	284.4	302.2	314.3	317.5

Fig. A8: Fuel Price Indices for the Industrial Sector, 1970 to 1997, UK

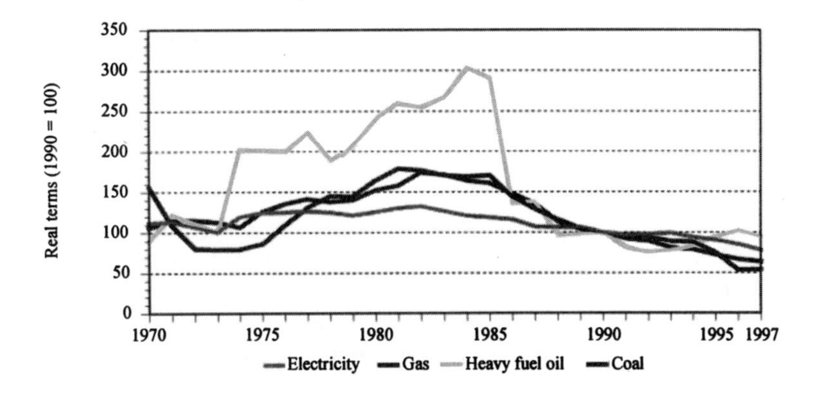

Tab. A21: Real Prices, 1990 = 100, UK

	1970	1980	1990	1995	1996	1997
Electricity	111.8	125.2	100.0	91.0	85.3	78.4
Gas	155.7	164.4	100.0	75.4	53.6	53.8
Heavy fuel oil	89.3	241.8	100.0	95.0	101.7	94.9
Coal	106.0	152.2	100.0	72.5	66.9	63.6
Industrial prices	116.9	150.7	100.0	87.7	80.6	75.3

Fig. A9: Fuel Price Indices for the Domestic Sector, 1970 to 1997, UK

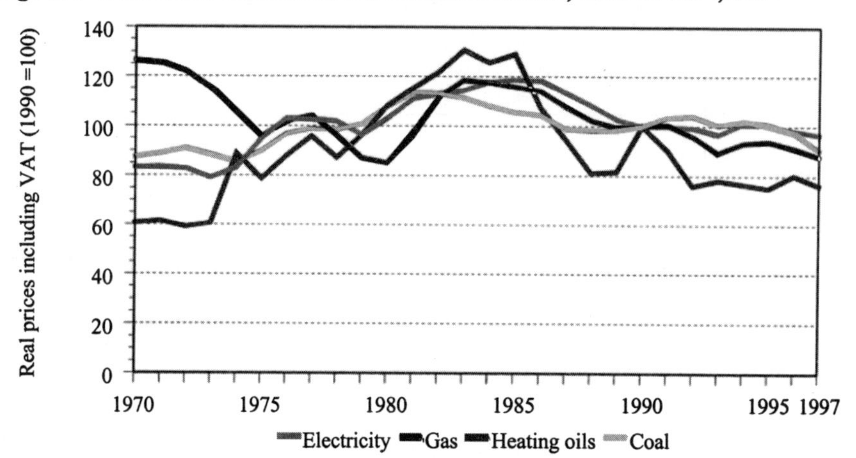

Tab. A22: Real Prices Including VAT, 1990 = 100, UK

	1970	1980	1990	1995	1996	1997
Electricity	83.2	102.8	100.0	100.9	97.4	90.4
Gas	126.6	85.1	100.0	93.9	91.2	88.0
Heating oils	60.6	108.1	100.0	75.1	80.2	76.2
Coal	87.4	107.7	100.0	100.4	98.3	96.6
Domestic prices (fuel & light)	89.1	97.6	100.0	96.9	94.2	89.0

Fig. A10: Carbon Dioxide Emissions by Source, 1970 to 1997, UK

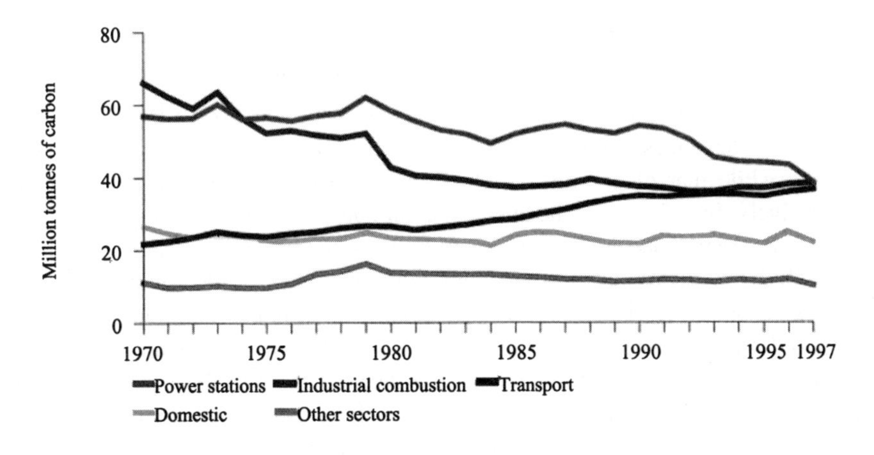

Tab. A23: Carbon Dioxide Emissions by Source, 1970 to 1997 (Mill. Tonnes of Carbon), UK

	1970	1980	1990	1995	1996	1997
Power stations	57	58	54	44	43	40
Industrial combustion	66	43	37	37	38	37
Domestic	26	23	22	22	25	23
Transport	22	26	35	35	36	37
Other sectors	11	14	11	11	12	12
Total	182	164	159	149	154	148

Source: Department of the Environment, Transport and the Regions

Box A1: Basic Aspects of CO_2, Energy Taxation and Innovations in Germany

Development of CO_2 Emissions in Germany

In Germany CO_2 emissions were strongly reduced in industry which has cut – within a long term shrinking process – its CO_2 emissions by 26.1% in the period 1990-97. The emissions of power generators and district heating firms fell by 14.9%, while that of private households and traffic increased by 8.6 and 6.4%, respectively. Small users in the business community reduced emissions by 26.3% in the same period; total emissions fell from 1014 tons to 894 tons. In 1997 power generation and district heating accounted for a share in CO_2 emissions of 37.3 and 19.9%, respectively. Transportation and private households represented a share of 20.5 and 15.5%, respectively (small users 6.3%). It will be difficult to reduce the emissions from private households as the demand for dwelling space and hence heating requirements are increasing in the long run. Even more difficult is transportation since both the numbers of cars and trucks is expected to increase in Germany and the whole EU in the future. Improved insulation of new houses and technological progress in the automotive industry as well as some incentives to shorten the length of average hauls in the transportation industry are potential ways to cut emissions in households and in transportation. It remains an open question how strong emissions in industry and power generation could be further reduced in Germany and the EU in the very long run.

Stage II of Ecological Tax Reform

On August 25, 1999 the German government decided in favour of a further future stage of the ecological tax reform. In the period 2000-2003 mineral oil taxes will be raised by 6 Pfennig (0.03 Euro) p.a. and taxes on electricity will be increased by 0.5 Pfennig per kWh. The revenue will be fully used to reduce social security contributions by one percentage point until 2003 (2000 by 0.1 percentage points, 2001-2003 0.3 percentage point p.a.). As of 2003 the overall tax burden per GJ and per t CO_2 will differ strongly across primary energy sources (see table).

Innovations and Environmental Progress

Since many years it has been recognized that innovations could play a crucial part within overall environmental policy (e.g. DOWNING/WHITE, 1986; WALLACE, 1995; BLAZEJCZAK/EDLER/HEMMELSKAMP/JÄNICKE, 1999, especially in the form of environmental-friendly (low-energy) product innovations and energy-saving or resource-saving process innovations. Some countries have implicitly incorporated innovation pressure in energy policies, e.g. in the 1990s Denmark with incentives for CO_2 efficient refrigerators or the UK in its 1989 Electricity Act which included non-fossil-fuel obligations and "renewable orders" (the regional distribution companies were compensated for higher electricity prices from non-fossil-fuel obligations by payments made out of the fossil-fuel-levy). It is obvious that raising prices of energy and energy-intensive products will stimulate certain innovations, which could bring about dynamic efficiency gains. There also is little doubt that environmental taxes could be both an element of internalizing negative external effects of pollution and an increasingly important element of the overall tax system (EEA, 1996; OECD, 1997). However, the concept of an innovation oriented ecological tax reform is the first approach for directly combining higher environmental taxes designed to internalize negative external effects with higher government R&D promotion aimed at internalizing positive external effects from innovations yielding better and cheaper products – a concept which could both improve allocation efficiency in a static and dynamic sense and stimulate output growth; supporting growth in turn should help to bring about sustained employment effects from reduced payroll/social security taxes. Thus the risks are minimized that price increases associated with higher energy/ CO_2 taxes and an intensified struggle for income – emerging in a situation of depressed growth – will result in renewed wage pressure which undermine initial employment gains. R&D promotion could focus both on generally raising private R&D expenditures, on creating more high-technology firms [with a high share focussing on environmentally relevant innovations] and on accelerating diffusion of new low-emission technologies and novel products of high-energy efficiency.

Tab. A24: Energy Taxation in Germany

	Unit (U)	Tax rate as of 1.4.1999	Increase 1999-2003				Total tax burden 2003		
			Total (cumulated)			at 1.4.99			
		Pf/U	Pf/U	DM/GJ	DM/t CO_2	Pf/U	Pf/U	DM/GJ	DM/CO_2
Coal	Kg	-	-	-	-	-	-	-	-
Heating oil (heavy), Heating	Kg	3,00	-	-	-	-	3,50	0,85	11
Heating oil (heavy), Power generation	Kg	5,00	-	-	-	-	3,50	0,85	11
Heating oil	L	8,00	4,00	1,12	15	4,00	12,00	3,37	46
Natural gas	KWh	0,36	0,32	0,89	16	0,32	0,68	1,89	34
Electricity	KWh	-	4,00	11,11	71	2,00	4,00	11,11	71
Fuel	L	98,00	30,00	9,27	129	6,00	128,00	39,55	549
Diesel	L	62,00	30,00	8,38	113	6,00	92,00	25,71	347

Source: DIW Wochenbericht 36/99 forthcoming

Tab. A25: EU Member State Commitments

Member States	Commitments in accordance with the Kyoto Protocol
Belgium	-7,5%
Denmark	-21%
Germany	-21%
Greece	+25%
Spain	+15%
France	0%
Ireland	+13%
Italy	-6,5%
Luxembourg	-28%
Netherlands	-6%
Austria	-13%
Portugal	+27%
Finland	0%
Sweden	+4%
United Kingdom	-12,5%
EU	-8%

Source: Appendix 1 in EU (1998). Community Strategy on Climate Change-Council Conclusions,Communiqué from the 2106 Th Council Meeting 16-17 June 1998, Luxembourg

Tab. A26: Benchmark Data for 1990

A. Summary statistics								
	GE	DK	SP	IT	FR	UK	REU	EU
GDP (% of EU total)	24	2	6	15	19	16	18	100
CO_2 (% of EU total)	32	2	7	13	12	18	17	100
CO_2 (% ton per capita)	13	11	5	7	7	10	8	9
Electricity (ton of CO_2/GWh)	653	892	429	565	105	686	260	462
B. CO_2 emissions sources								
	GE	DK	SP	IT	FR	UK	REU	EU
Electricity	36	43	31	30	12	38	23	30
Energy production	3	2	6	5	5	5	5	4
Industry	20	11	21	20	23	14	24	20
Transport	17	26	32	25	34	24	26	24
Final demand	24	17	10	20	26	19	22	21
Total	100	100	100	100	100	100	100	100

Note: The Rest of the European Union (REU) includes Austria, Belgium, Finland, Luxembourg, Greece, Ireland, Netherlands, Portugal and Sweden

Source: Böhringer, C.; Jensen, J.; Rutherford, T. F. (1999), Energy Market Projections and Differentiated Carbon Abatement in the European Union, Discussion paper N. 99-11, ZEW, Mannheim, p. 14

Bibliography

ACS, Z., AUDRETSCH, D., FELDMAN, M. (1994), R&D Spillovers and Recipient Firm Size, in: The Review of Economics and Statistics, 76, pp. 336–340.

ADAMS, J. (1990), Fundamental Stocks of Knowledge and Productivity Growth, in: Journal of Political Economy, 98, pp. 673 – 702.

ADDISON, J. und WELFENS, P.J.J., eds. (1999), Labor Markets and Social Security, 2nd revised and enlarged edition, Heidelberg and New York: Springer.

ADDISON, J. und WELFENS, P.J.J., eds. (1998), Labor Markets and Social Security, Heidelberg and New York: Springer.

ALMON, C. (1991), The Inforum Approach to Interindustry Modelling. Economic Systems Research 3, 1-7.

ALTNER, G., DÜRR, H. P., MICHELSEN, G., NITSCH, J. (1995), Zukünftige Energiepolitik – Vorrang für rationelle Energienutzung und regenerative Energiequellen, Bonn.

AMANO, A. (1998), Climate Change, Response Timing, and Integrated Assessment Modeling, in: Environmental Economics and Policy Studies, 1/1, pp.3-18.

ARNDT, H.-W., HEINS, B., HILLEBRAND, B., MEYER, E.C., PFAFFENBERGER, W., STRÖBELE, W. (1998), Ökosteuern auf dem Prüfstand der Nachhaltigkeit, Berlin.

BARDY, R. (1974), Die Produktivität von Forschung und Entwicklung, Meisenheim am Glan.

BARTELSMAN, E., VAN LEEUWEN, G., NIEUWENHUIJSEN, H., ZEELENBERG, K. (1996), R&D and Productivity Growth: Evidence from Firm-Level Data in the Netherlands, Paper presented at the Conference of the European Economic Association, Istanbul.

BAYOUMI, T., COE, D.T., HELPMAN, E. (1999), R&D spillovers and global growth, in: Journal of International Economics, 47, pp. 399 – 428.

BLAZEJCZAK, J., EDLER, D., HEMMESKAMP, J., JÄNICKE, M., Umweltpolitik und Innovation: Politikmuster und Innovationswirkungen im internationalen Vergleich, Zeitschrift für Umweltpolitik und Umweltrecht, Vol. 22, pp. 1-32.

BMWi (1999), Energie Daten 1999, Bonn, p. 40.

BÖNTE, W. (1998), Wie produktiv sind Investitionen in Forschung und Entwicklung, Diskussionspapier, Institut für Allokation und Wettbewerb, Universität Hamburg.

BRÄNNLUND, R. (1999), Green Tax Reforms: Some Experiences from Sweden, in: SCHLEGELMILCH, K., ed., Green Budget Reform in Europe, Heidelberg and New York: Springer.

BUDD, A. und HOBBIS, S. (1989), Cointegration, Technology and the Long-Run Production Function, Discussion Paper, London Business School Centre for Economic Forecasting.

BUDD, A und HOBBIS, S. (1989a), Output Growth and Measure of Technology, Discussion Paper, London Business School Centre for Economic Forecasting.

CAMERON, C. (1998), Innovation and Growth: A Survey of the Empirical Evidence, Working Paper, Nuffield College, Oxford.

COE, D.T., HELPMAN, E. (1995), International R&D spillovers, in: European Economic Review, 39, pp.859-887.

COE, D.T. und MOGHADAM, R. (1993), Capital and Trade as Engines of Growth in France, in: IMF Staff Papers, 40, pp. 542–566.

COHENDET, P., LLERENA, P., SORGE, A. (1992), Technological Diversity and Coherence in Europe: An Analytical Overview, Revue d'Economie Industrielle, 59, pp. 9-26.

CONRAD, K. (1995), Choosing Emission Taxes under International Price Competition, Fondazione Enrico Mattei, Milano, mimeo.

CUNEO, P. und MAIRESSE, J. (1984), Productivity and R&D at the Firm Level in French Manufacturing, in: GRILICHES, Z. (ed.), R&D, Patents and Productivity, Chicago.

DEUTSCHE BANK RESEARCH (1999), Kernenergie: Ist ein Ausstieg möglich?, Sonderbericht, Frankfurt.

DOSI, G. (1982), Technological Paradigms and Technological Trajectories, A Suggested Interpretation of the Determinants and Directions of Technical Change, in: Research Policy, 11, pp. 147-162.

DOSI, G., PAVITT, K., SOETE, L. (1990), The Economics of Technical Change and International Trade, New York etc.

DOWNING, P. und WHITE, L.J., Innovation in Pollution Control", Journal of Environmental Economics and Management, No. 13, 886, pp. 18-29.

EEA (European Environmental Agency) (1996), Environmental Taxes. Implementation and Environmental Effectiveness, Copenhagen.

ENGELBRECHT, H.J. (1998), Business Sector R&D and Australia's Manufacturing Trade Structure, in: Applied Economics, 30, pp. 177-187.

EUROPEAN COMMISSION (1999), Annual Economic Report, Brussels.

EUROSTAT (1999), Electricity Statistics, Theme 8 –01/1999, Luxembourg.

FAGERBERG, J. (1988), International Competitiveness, in: Economic Journal, 98, pp. 355 – 374.

GALE, R., BARG, S., GILLIES, A. (1995), eds. Green Budget Reform, Earthscan,London.

GEHRKE, B. und GRUPP, H. (1994), Hochtechnologie und Innovationspotential, Heidelberg.

GREENPEACE (1997), Energy Subsidies in Europe, Amsterdam.

GRILICHES, Z. (1995), R&D and Productivity: Econometric Results and Measurement Issues, in: STONEMAN, P. (ed.), Handbook of the Economics of Innovation and Technological Change, Oxford, UK/Cambridge, MA, pp. 52–89.

GRILICHES, Z. und LICHTENBERG, F. (1984), R&D and Productivity Growth at the Industry Level: Is There still a Relationship, in: GRILICHES, Z. (ed.), R&D, Patents and Productivity, Chicago, pp. 465-496.

GRILICHES, Z. und MAIRESSE, J. (1983), Comparing Productivity Growth: An Exploration of French and US Industrial and Firm Data, European Economic Review, 21, pp. 89–119.

GROENEWOLD, N. (1999), Employment Protection and Aggregate Unemployment, Journal of Macroeconomics, Vol. 21, pp. 619-630.

GROSSMAN, G. und HELPMAN, E. (1991), Innovation and Growth in a Global Economy, Cambridge, Mass.

GRUPP, H. und JUNGMITTAG, A. (1999), Convergence in Global High Technology? A Decomposition and Specialisation Analysis for Advanced Countries, in: Jahrbücher für Nationalökonomie und Statistik, 218, pp. 552–573.

HALL, B. und MAIRESSE, J. (1995), Exploring the Relationship between R&D and Productivity in French Manufacturing Firms, in: Journal of Econometrics, 65, pp. 263 – 294.

HENSING, I. (1996), Die Perspektive von Kernenergie in wettbewerblich geöffneten Energiemärkten, in: Zeitschrift für Energiewirtschaft, pp. 53-64.

HENSING, I., PFAFFENBERGER, W., STRÖBELE, W. (1998), Energiewirtschaft, München.

HENSING, I. und SCHULZ, W. (1995), Wirtschaftlichkeitsvergleich verschiedener Entsorgungspfade von Kernkraftwerken - eine Kostensimulation aus deutscher Sicht, München.

HODGSON, G.M. (1993), Economics and Evolution, Bringing Life back to Economics, Cambridge.

INTERNATIONALE ENERGIE-AGENTUR (1998), CO_2 Emissions from fuel combustion, edition, Paris.

JAFFE, A.B., TRAJTENBERG, M., HENDERSON, R. (1992), Geographic Localization of Knowledge Spillovers as Evidence by Patent Citations, NBER Working Paper, 3993, Cambridge, Mass.

JASINSKI, P. und PFAFFENBERGER, W., eds. (1999), Energy Liberalisation and the Environment: Multi-Regulation in Eastern and Western Europe, Aldershot, Ashgate, forthcoming.

JASINSKI, P. und SKOCZNY, T., eds. (1996a), Studia nad integracją europejską: Elektroenergetyka [European Integration Studies: The Electricity Supply Industry], RPRC, Oxford and Centre for Europe, Warsaw.

JASINSKI, P. und SKOCZNY, T., eds. (1996b), Studia nad integracją europejską: Gazownictwo [European Integration Studies: The Gas Industry], RPRC, Oxford and Centre for Europe, Warsaw.

JONES, C.I. und WILLIAMS, J.C. (1997), Measuring the Social Return to R&D, Working Paper, Stanford University.

JUNGMITTAG, A., BLIND, K., GRUPP, H. (1999), Innovation, Standardisation and the Long-term Production Function, in: Zeitschrift für Wirtschafts- und Sozialwissenschaften, 119, pp. 209–226.

JUNGMITTAG, A., GRUPP, H., HINZE, S., HULLMANN, A., SCHMOCH, U. (1998), Berichterstattung zur technologischen Leistungsfähigkeit Deutschlands 1997, Materialien des Fraunhofer-Instituts für Systemtechnik und Innovationsforschung, Karlsruhe.

JUNGMITTAG, A., GRUPP, H., HULLMANN, A. (1998), Changing Patterns of Specialisation in Global High Technology Markets: An Empirical Investigation of Advanced Countries, in: Vierteljahreshefte zur Wirtschaftsforschung, 67, 1, pp 86 – 98.

JUNGMITTAG, A., MEYER-KRAHMER, F., REGER, G. (1999), Globalisation of R&D and Technology Markets – Trends, Motives, Consequences, in: MEYER-KRAHMER, F., ed., Globalisation of R&D and Technology Markets – Consequences for National Innovation Policies, Heidelberg/New York, pp. 37 – 77.

JUNGMITTAG, A. und WELFENS, P.J.J. (1996), Telekommunikation, Innovation und die langfristige Produktionsfunktion: Theoretische Aspekte und eine Kointegrationsanalyse für die Bundesrepublik Deutschland, Diskussionsbeitrag 20 des Europäischen Instituts für Internationale Wirtschaftsbeziehungen (EIIW), Potsdam.

KALTSCHMITT, M. und WIESE, A., Hrsg. (1995), Erneuerbare Energien - Systemtechnik, Wirtschaftlichkeit, Umweltaspekte, Berlin, Heidelberg.

KITSCHELT, H. (1980), Kernenergiepolitik – Arena eines gesellschaftlichen Konflikts, Frankfurt.

LOEFFELHOLZ, H.D. Von (1999), Steuerreform: Erfordernisse, Spielräume, Wirkungen, RWI Mitteilungen Vol. 49, pp. 161-173.

LOSKE, R. (1996), Klimapolitik, Marburg: Metropolis.

LUCAS Jr., R.E. (1988), On the mechanics of economic development, in: Journal for Monetary Economics, 22, pp. 3-42.

MA, Q. (1997), A Bilateral Trade Model for the Inforum International System, in: Tomaczewicz, L. (ed.), Proceedings of the 3rd World Inforum Conference, Lodz.

MACKSCHEIDT, K. (1996), Die ökologische Steuerreform im Lichte steuerpolitischer Ideale, in: KÖHN, J. und WELFENS, M., eds., Neue Ansätze in der Umweltökonomie, Marburg: Metropolis, pp. 109-125.

MAIRESSE, J. und CUNEO, P. (1985), Recherche-développement et performances des entreprises: Une étude économétrique sur des données individuelles, in: Revue Économique, 36, pp. 1001 – 1042.

MAIRESSE, J. und HALL, B. (1996), Estimating the Productivity of Research and Development in French and United States Manufacturing Firms: An Exploration of Simultaneity Issues with GMM Methods, in: WAGNER, K., van ARK, B., eds., International Productivity Differences: Measurement and Explanations, Amsterdam.

MARKEWITZ, P. und MARTINSEN, D. (1999), Kernenergie ohne zielorientierte CO_2-Minderungsstrategie, in: Energiewirtschaftliche Tagesfragen, pp. 60–63.

MEYER, B. und EWERHART, G. (1997), Lohnsatz, Produktivität und Beschäftigung. Ergebnisse einer Simulationsstudie mit dem disaggregierten ökonometrischen Modell Inforge, in: SCHNABL, H. (ed.), Innovation und Arbeit: Fakten- Analyse- Perspektiven, Mohr, Tübingen.

MEYER, B. und EWERHART, G. (1999), Inforge. Ein disaggregiertes Simulations- und Prognosemodell für Deutschland, in: LORENZ, H.W. und MEYER, B. (eds.), Studien zur Evolutorischen Ökonomik IV, Berlin: Duncker und Humblot.

MEYER, B. und WELFENS, P.J.J. (1999), Innovation-Augmented Ecological Tax Reform: Theory, Model Simulation and New Policy Implications, EIIW Discussion Paper, 65, University of Potsdam.

MEYER, B., BOCKERMANN, A., EWERHART, G., LUTZ, C. (1999), Marktkonforme Umweltpolitik. Wirkungen auf Luftschadstoffemissionen, Wachstum und Struktur der Wirtschaft, Heidelberg: Physica.

MEYER-KRAHMER, F. (1992), The Effects of New Technologies on Employment, in: Economics of Innovation and New Technology, 2, pp. 131 – 149.

MEYER-KRAHMER, F. und WESSELS, H. (1989), Intersektorale Verflechtung von Technologiegebern und Technologienehmern. Eine empirische Analyse für die Bundesrepublik Deutschland, Jahrbücher für Nationalökonomie und Statistik, Vol 206/6.

MICHAELIS, H. und SALANDER, C. (1995), Handbuch der Kernenergie, 4. Auflage, Frankfurt.

MÖHNEN, P., NADIRI, M., PRUCHA, I. (1986), R&D, Production Structure and Rates of Return in the US, Japanese and German Manufacturing Sectors, in: European Economic Review, 30, pp. 749–771.

NITSCH, J. (1995), Potentiale und Märkte der Kraft-Wärme-Kopplung in Deutschland, DLR Institut für Technische Thermodynamik, STB Bericht, 15, Stuttgart.

NITSCH, J. (1999), Erneuerbare Energie an der Schwelle zum nächsten Jahrtausend - Rückblick und Perspektiven, Mskr.

NORDHAUS, W. D. (1997), The Swedish Nuclear Dilemma, Washington.

NYHUS, D. (1991), The Inforum International System. Economic Systems Research, 3, pp. 55-64.

NYHUS, D. und WANG, Q. (1997), Investments and Exports: A Trade Share Perspective, Paper presented at the 5. World Inforum Conference, Bertinoro.

O'MAHONY, M. und WAGNER, K. (1996), Changing Fortune: An Industry Study of British and German Productivity Growth over Three Decades, in: MAYES, D. (ed.), Sources of Productivity Growth in the 1980s, Cambridge.

OATES (1995), Green Taxes: Can We Protect the Environment and Improve the Tax System at the same Time?, in: Southern Economic Journal, 61 (4).

OECD (1997), Environmental and Green Tax Reform, Paris.

OECD (1998), Energy Policies of IEA Countries. Germany 1998 Review, Paris.

OTTE, C. und PFAFFENBERGER, W. (1999), Energieeffizienz in Deutschland, in: PFAFFENBERGER, W. und STREBEL, H. (Hrsg.), Ökonomische Energienutzung, München, pp. 75-126.

PATEL, P. und SOETE, L. (1988), L'Évaluation des effets économiques de la technologie, in: STI Review, 4, pp. 133–183.

PERNER, J. und RIECHMANN, C. (1998), Netzzugang oder Durchleitung?, in: Zeitschrift für Energiewirtschaft, pp. 41-57.

PFAFFENBERGER, W. (1995), Arbeitsplatzeffekte von Energiesystemen, Frankfurt.

PFAFFENBERGER, W. (1999), Ausstieg aus der Kernenergie – und was kommt danach?, Frankfurt/München: Piper.

PFAFFENBERGER, W. und KEMFERT, C. (1998), Beschäftigungseffekte durch eine verstärkte Nutzung erneuerbarer Energien, Bonn.

PROGNOS-GUTACHTEN (1995), Die Energiemärkte Deutschlands im zusammenwachsenden Europa - Perspektiven bis zum Jahr 2020, Stuttgart.

ROMER, P.M. (1986), Increasing Returns and Long-Run Growth, in: Journal of Political Economy, 94, pp. 1002–1037.

ROMER, P.M. (1990), Endogenous Technological Change, in: Journal of Political Economy, 98, pp. 71-102.

ROTHFELS, J. (1998), Umweltschutz und internationale Wettbewerbsfähigkeit aus Sicht der neuen Außenhandelstheorie, in: HORBACH, J., MEIßNER, T, ROTHFELS, J., HOLST, K., VOIGT, P., Umweltschutz und Wettbewerbsfähigkeit, Baden-Baden: Nomos, pp. 15-33.

ROTHFELS, J., Umweltpolitik und unternehmerische Anpassung, forthcoming.

RWI (1990), Stellungnahme zum Entwurf eines Gesetzes zum Einstieg in die ökologische Steuerreform, Bundestagsanhörung vom 18.1.1999, RWI, Essen.

SANDMO, A. (1975), Optimal Taxation in the presence of Externalities, in: Swedish Journal of Economics, 77 (1).

SAVIOTTI, P.P. (1990), The Role of Variety in Economic and Technological Development, in: SAVIOTTI, P.P. und METCALFE, J.S. (eds.), Evolutionary Theories of Economic and Technological Change: Present Status and Future Prospects, Reading.

SCHERER, F. (1982), Inter-Industry Technology Flow and Productivity Growth, in: Review of Economics and Statistics, 64.

SCHLEGELMILCH, K., ed. (1999), Green Budget Reform in Europe, Heidelberg and New York.

SCHMIDT, T.F.N. und KOSCHEL, H. (1999), GEM-E3, in: FAHL, U. und LÄGE, E. (ed.), Strukturelle und gesamtwirtschaftliche Auswirkungen des Klimaschutzes: Die nationale Perspektive, Heidelberg.

STONEMAN, P., ed., Handbook of the Economics of Innovation and Technological Change, Oxford.

VAN DEN BERGH, J.C.J.M. and VAN DER STRAATEN, J. (1994), Toward an Ecological Tax Reform.

WAKELIN, K. (1998), The role of innovation in bilateral OECD trade performance, in: Applied Economics, 30, 1335-1346.

WALLACE, D. (1995), Environmental Policy and Industrial Innovation: Strategies in Europe, the USA and Japan, London.

WELFENS, P.J.J. (1999a), Globalization of the Economy, Unemployment and Innovation, Heidelberg and New York.

WELFENS, P.J.J. (1999b), Beschäftigungsfördernde Steuerreform in Deutschland zum Euro-Start: Für eine wachstumsorientierte Doppelsteuerreform, RWI-Mitteilungen, Vol. 49, 149-160, original version published as EIIW-Analysen zur Wirtschaftspolitik No.3 (see http://www.euroeiiw.de), October 7, 1998, University of Potsdam.

WELFENS, P.J.J. (1999c), EU Eastern Enlargement and the Russian Transformation Crisis, Heidelberg and New York.

WELFENS, P.J.J. und YARROW, G., eds. (1997), Telecommunications and Energy in Systemic Transformation, Heidelberg and New York.

WELFENS, P.J.J., AUDRETSCH, D., ADDISON, J.,GRUPP, H. (1998), Technological Competition, Employment and Innovation Policies in OECD Countrys, Heidelberg and New York.

WELFENS, P.J.J., GRAACK, C., GRINBERG, R. YARROW, G., eds. (1999), Towards Competition in Network Industries, New York.

WELSCH, H. (1999), Lean, in: FAHL, U. und LÄGE, E., eds., Strukturelle und gesamtwirtschaftliche Auswirkungen des Klimaschutzes: Die nationale Perspektive, Heidelberg.

WITT, U., ed. (1993), Evolutionary Economics, The International Library of Critical Writings in Economics, 25, Aldershot.

WUPPERTAL INSTITUTE FOR CLIMATE (1998), Energy Pricing Policy: Targets, Possibilities and Impacts, Energy and Research Series, ENER102, 2-1998, European Parliament, Luxembourg.

List of Tables

List of Figures

Further Publications by *Paul J. J. Welfens*

P. J. J. Welfens
Market-oriented Systemic Transformations in Eastern Europe
Problems, Theoretical Issues, and Policy Options
1992. XII, 261 pp. 20 figs., 29 tabs.,
Hardcover, ISBN 3-540-55793-8

M. W. Klein, P. J. J. Welfens
Multinationals in the New Europe and Global Trade
1992. XV, 281 pp. 24 figs., 75 tabs.,
Hardcover, ISBN 3-540-54634-0

R. Tilly, P. J. J. Welfens
European Economic Integration as a Challenge to Industry and Government
Contemporary and Historical Perspectives on International Economic Dynamics
1996. X, 558 pp. 43 figs., Hardcover,
ISBN 3-540-60431-6

P. J. J. Welfens
European Monetary Integration
EMS Developments and International Post-Maastricht Perspectives
3rd revised and enlarged edition
1996. XVIII, 384 pp. 14 figs., 26 tabs.,
Hardcover, ISBN 3-540-60260-7

P. J. J. Welfens
European Monetary Union
Transition, International Impact and Policy Options
1997. X, 467 pp. 50 figs., 31 tabs.,
Hardcover, ISBN 3-540-63309-7

P. J. J. Welfens, G. Yarrow
Telecommunications and Energy in Systemic Transformation
International Dynamics, Deregulation and Adjustment in Network Industries
1997. XII, 501 pp. 39 figs., Hardcover,
ISBN 3-540-61586-5

P. J. J. Welfens, H. C. Wolf
Banking, International Capital Flows and Growth in Europe
Financial Markets, Savings and Monetary Integration in a World with Uncertain Convergence
1997. XIV, 458 pp. 22 figs., 63 tabs.,
Hardcover, ISBN 3-540-63192-5

P. J. J. Welfens
Economic Aspects of German Unification
Expectations, Transition Dynamics and International Perspectives
2nd revised and enlarged edition
1996. XV, 527 pp. 34 figs., 110 tabs.,
Hardcover, ISBN 3-540-60261-5

P. J. J. Welfens, D. Audretsch, J. T. Addison and H. Grupp
Technological Competition, Employment and Innovation Policies in OECD Countries
1998. VI, 231 pp. 16 figs., 20 tabs.,
Hardcover, ISBN 3-540-63439-8

J. T. Addison, P. J. J. Welfens
Labor Markets and Social Security
Wage Costs, Social Security Financing and Labor Market Reforms in Europe
1998. IX, 404 pp. 39 figs., 40 tabs.,
Hardcover, ISBN 3-540-63784-2

P. J. J. Welfens
EU Eastern Enlargement and the Russian Transformation Crisis
1999. X, 151 pp. 12 figs., 25 tabs.,
Hardcover, ISBN 3-540-65862-9

P. J. J. Welfens
Globalization of the Economy, Unemployment and Innovation
1999. VI, 255 pp. 11 figs., 31 tabs.,
Hardcover, ISBN 3-540-65250-7

P. J. J. Welfens, G. Yarrow, R. Grinberg, C. Graack
Towards Competition in Network Industries
Telecommunications, Energy and Transportation in Europe and Russia
1999. XXII, 570 pp. 63 figs., 63 tabs.,
Hardcover, ISBN 3-540-65859-9

P. J. J. Welfens, J. T. Addison, D. B. Audretsch, T. Gries, H. Grupp
Globalization, Economic Growth and Innovation Dynamics
1999. X, 160 pp. 15 figs., 15 tabs.,
Hardcover, ISBN 3-540-65858-0

R. Tilly, P.J.J. Welfens
Economic Globalization, International Organizations and Crisis Management
Contemporary and Historical Perspectives on Growth, Impact and Evolution of Major Organizations in an Interdependent World
2000. XII, 408 pp. 11 figs., 20 tabs.,
Hardcover, ISBN 3-540-65863-7

P.J.J. Welfens, E. Gavrilenkov
Restructuring, Stabilizing and Modernizing the New Russia
Economic and Institutional Issues
2000. XIV, 516 pp. 82 figs., 70 tabs.,
Hardcover, ISBN 3-540-67429-2

P.J.J. Welfens
European Monetary Union and Exchange Rate Dynamics
New Approaches and Applications to the Euro
2001. X, 159 pp. 26 figs., 12 tabs.,
Hardcover, ISBN 3-540-67914-6

P.J.J. Welfens, B. Meyer, W. Pfaffenberger, P. Jasinski, A. Jungmittag
Energy Policies in the European Union
Germany's Ecological Tax Reform
2001. VIII, 143 pp. 21 figs., 41 tabs.,
Hardcover, ISBN 3-540-41652-8

Recent Titles in Economics

P.J.J. Welfens

Globalization of the Economy, Unemployment and Innovation

Structural Change, Schumpetrian Adjustment, and New Policy Challenges

Economic globalization has intensified since the 1980s and created faster channels of international interdependence and an accelerating technology race. In this new asymmetric world economy, the EU is facing a dynamic and flexible US system which takes advantage of the global quest for foreign direct investment. Innovation policies in the EU - in particular in Germany - are found to be rather inadequate. There are also new theoretical challenges where a „structural macro model" and a Schumpetrian model of innovation and full employment are presented as new approaches. Besides theoretical challenges, the increasing global dynamics raise new problems of international policy coordination which could lead to unsustainable economic globalization.

1999. VI, 255 pp. 11 figs., 31 tabs.
Hardcover DM 119* / £ 46 / FF 449 / Lit. 131.420 / sFr 103,-
ISBN 3-540-65250-7

P.J.J. Welfens, G. Yarrow, R. Grinberg, C. Graack (Eds.)

Towards Competition in Network Industries

Telecommunications, Energy and Transportation in Europe and Russia

1999. XXII, 570 pp. 63 figs., 63 tabs.
Hardcover DM 179* / £ 69 / FF 675 / Lit. 197.690 / sFr 154,-
ISBN 3-540-65859-9

P.J.J. Welfens

EU Eastern Enlargement and the Russian Transformation Crisis

1999. X, 151 pp. 12 figs., 25 tabs.
Hardcover DM 98* / £ 3.,50 / FF 370 / Lit. 108.230 / sFr 86,50
ISBN 3-540-65862-9

Please order from
Springer-Verlag
P.O. Box 14 02 01
D-14302 Berlin, Germany
Fax: +49 30 827 87 301
e-mail: orders@springer.de
or through your bookseller

* Recommended retail prices. Prices and other details are subject to change without notice.
In EU countries the local VAT is effective. d&p · BA 41652/1 SF

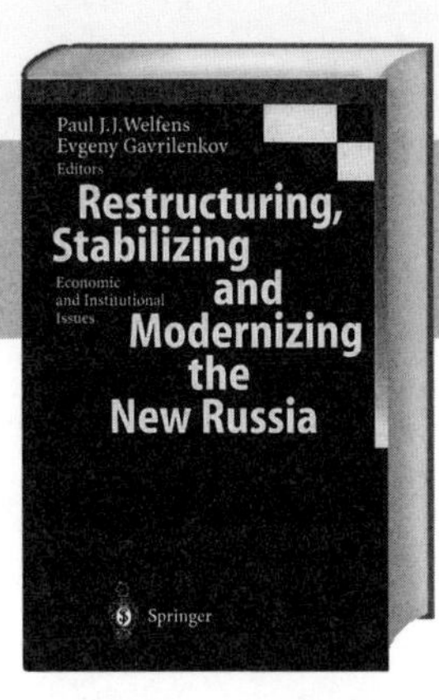

P.J.J. Welfens, E. Gavrilenkov, (Eds.)

Restructuring, Stabilizing and Modernizing the New Russia

Economic and Institutional Issues

The Russian transformation crisis is analyzed and put into the context of national policy failures and international shocks. The study also has a focus on structural change and foreign direct investment - including an econometric perspective. Basic problems of economic opening-up, macroeconomic policy and aspects of political economy are also discussed. The role of institutional reforms, membership in international organizations and supply-side policies are emphasized.

2000. XIII, 516 pp. 82 figs., 70 tabs. Hardcover * **DM 179**; £ 61.50; FF 675; Lit. 197.690; sFr 154 ISBN 3-540-67429-2

M. Casson, A. Godley, (Eds.)

Cultural Factors in Economic Growth

This volume incorporates contributions of scholars from economics, management studies and international relations, as well as economic and social historians' attempts to evaluate the role and impact of cultural factors on economic growth.

2000. VIII, 244 pp. 2 figs., 11 tabs. (Studies in Economic Ethics and Philosophy) Hardcover * **DM 139**; £ 48; FF 524; Lit. 153.520; sFr 120 ISBN 3-540-66293-6

C. Yoshida

Illegal Immigration and Economic Welfare

Illegal immigration is a problem to not only a labor importing country but also to a labor exporting country, since the implementation of strict immigration policies, i.e., border patrol and employer sanctions affects both economies. The purpose of this book is to complement previous studies on deportable aliens. The effects of such enforcement policies on the income or welfare of the foreign (labor exporting) country, the home (labor importing) country, and the combined (global) income of the two countries are examined.

2000. XIV, 158 pp., 1 fig., 2 tabs. (Contributions to Economics) Softcover * **DM 75**; £ 26; FF 283; Lit. 82.820 sFr 66 ISBN 3-7908-1315-X

T. Steger

Transitional Dynamics and Economic Growth in Developing Countries

Four stylised facts of aggregate economic growth are set up initially. The growth process is interpreted to represent transitional dynamics rather than balanced-growth equilibria. Against this background, the fundamental importance of subsistence consumption is comprehensively analysed. Subsequently, the meaning of the productive-consumption hypothesis for the intertemporal consumption trade-off and the growth process is investigated. Finally, the process of growth is analysed empirically by means of cross-sectional conditional convergence regressions with endogenous control variables.

2000. VIII, 151 pp. 21 figs. 5 tabs. (Lecture Notes in Economics and Mathematical Systems, Vol. 489) Softcover * **DM 72**; £ 25; FF 272; Lit. 79.510; sFr 63,50 ISBN 3-540-67563-9

Please order from
Springer · Customer Service
Haberstr. 7
69126 Heidelberg, Germany
Tel: +49 (0) 6221 - 345 - 217/8
Fax: +49 (0) 6221 - 345 - 229
e-mail: orders@springer.de
or through your bookseller

* Recommended retail prices. Prices and other details are subject to change without notice.
In EU countries the local VAT is effective. d&p · BA 41652/2 SF

Printing: Weihert-Druck GmbH, Darmstadt
Binding: Buchbinderei Schäffer, Grünstadt